NICOLE ORESME, *TRAITIÉ DE L'ESPERE*

TEXTES VERNACULAIRES DU MOYEN ÂGE

VOLUME 35

Founded in 2004, Textes Vernaculaires du Moyen Âge is an editorial enterprise designed to meet the needs of scholars and students alike. Its main focus is on texts which have hitherto failed to benefit from adequate editorial treatment and which, as a consequence, remain unknown or imperfectly known to the academic community. All aspects of medieval vernacular literary activity form part of its brief: literary texts, historical writings, including chronicles, devotional treatises, sermons, scientific treatises, and the like. The series also welcomes editions of better-known texts accompanied by modern translations, designed to meet the needs of scholars and students who may be unfamiliar with the language of the original texts. Each edition comprises a description of the source text or texts, manuscript or early printed book, accompanied by explanatory notes and a comprehensive glossary. All submissions are subject to blind or double blind peer review. The series concentrates on propositions in both medieval English and French. Potential editors are strongly advised to contact the general editor in the first instance before making a submission: Stephen.Morrison@univ-poitiers.fr.

Nicole Oresme, *Traitié de l'espere*

A critical edition

Edited by
RONALD B. THOMSON

BREPOLS

D/2025/0095/349
ISBN 978-2-503-62091-6
eISBN 978-2-503-62097-8
DOI 10.1484/M.TVMA-EB.5.151742
ISSN 1782-6233
eISSN 2566-0225

Printed in the EU on acid-free paper.

TABLE OF CONTENTS

NOTE: There is a list of Oresme's chapters (with starting line numbers) in the Prologue to the *Traitié* on pages 24-26.

The illustration on the cover is taken from the presentation copy of the *Traitié* (Oxford, St. John's College, ms. 164, fol. 1r) and shows King Charles V of France, Oresme's patron, examining a number of texts that he had requested to be translated into or written in Middle French. See below, p. 23, note 1.

PREFACE

To some degree Nicole Oresme has always been of interest to me. During my graduate studies I looked closely into his various writings and I still have those notes. At some point, but later I think, I came across a bound copy of Lillian McCarthy's thesis,[1] an edition of Oresme' *Traitié de l'espere*, which at that point seemed to be the only available modern edition of the text. I conceived of the idea of recasting the thesis by converting it into an up-to-date version which could be published (with updated notes), thus filling a gap in the modern Oresme corpus.

However, to do this I needed permission of McCarthy or her heirs (when I learned that she had passed away), and with some digging found a nephew who lived in Renfrew, Ontario. Several letters to him went unanswered, so I was investigating the possibility of applying to the Canadian Copyright Board for permissiont o reproduce McCarthy's text.

But on further examination of the thesis, and further research, I found that since the time of its preparation (in 1943) several more manuscripts of the text had been identified, including one (Oxford, St. Johns College, ms. 164) which seemed a better candidate for base text than the one McCarthy had used (Paris, Bibliothèque nationale de France, ms. fr. 1350). At this point it became obvious that a totally new edition was needed, not only in light of the other manuscripts, but in terms of more recent scholarship.

I also discovered that the text had been edited even earlier, by John Vernon Meyers, as a masters thesis[2] at the University of Syracuse in 1940, again using

---

1 Lillian Margaret McCarthy (1915-1991), "Maistre Nicole Oresme, Traitié de l'Espere," Doctor of Philosophy thesis, University of Toronto, 1943. At the moment there are 3 copies of this thesis in Toronto – one in the University of Toronto Archives with other doctoral dissertations, one in the Gerstein Science Information Centre, and one in the library of the Pontifical Institute of Mediaeval Studies. The thesis is not available through Proquest (as with other North American thesis), but there are 2 copies (on microfilm?) on deposit at Library and Archives Canada in Ottawa.

2 John Vernon Myers (1916-1974), "Maistre Nicole Oresme, 'Le Traité de la Sphère', prepared from the text of MS. Français 1350, Bibliothèque Nationale with variant notes and a critical introduction," Masters thesis, Department of Romance Languages, Syracuse University, 1940. It appears that the supervisor was Albert D. Menut who went on to edit (with Alexander J. Denomy) Oresme's *Le livre du ciel et du monde* (Madison: University of Wisconsin Press, 1975),

The thesis is shelved in the University of Syracuse Library, although I have been able to obtain a copy which is now available for consultation at the library of the Pontifical Institute of Mediaeval Studies in Toronto, next to McCarthy's thesis. Again, no copy seems to be available

Paris, BnF, ms fr. 1350 as the base text. Meyers himself is somewhat enigmatic; it appears that he disappeared from academia for the duration of World War II, but then completed a doctoral thesis in 1959, which was later published. It also appears that he did not hold a university position afterwards and apparently died in 1974, without publishing anything further. I found no record of a family.

Thirdly, the text was edited by Terrence Joseph McGovern based on the presentation copy only (Oxford, St. John's College, ms 164) as a doctoral thesis at the Catholic University of America in 1974. This version (and not a critical edition as indicated by the title) was available on-line and, although not without its problems, has proved useful in checking my own transcription. McGovern seems not to have known of the two other theses.

In light of these developments I have prepared this edition. It is aimed at both scholars of Middle French and of medieval science / astronomy. Because of this I have added some notes about Middle French which might help the historians of science understand this element; and notes about the scientific content which might help the scholars of Middle French. I have also tried to pin down all the source references that Oresme cites, many of which have not hitherto been identified – and link them to modern editions, if possible.

McCarthy's thesis contained one element which is not included here – a study of the dialects (mostly northern France) of the various manuscripts. Anyone interested in this specialized topic should consult her thesis. She also compiled a list of neologisms in French that Oresme was the first to introduce. This I have included (and updated the cross references to Old and Middle French dictionaries), eliminating any entries for which earlier instances have been discovered. Menut and Denomy also included such a list in their edition of *Le livre du ciel et du monde*, and I have combed their list to see if in fact any of those words were introduced in the earlier *Traité de l'espere.*

Further comments on the nature of the Middle French text and of this edition will be found in the introduction. However, I am unable to give a full linguistic analysis of Oresme's text – that is not my area of expertise. But I hope that by making this text available, it will enable others to do such an analysis, not only of the *Traité* but of the other French-language texts of Oresme as well.

As you will note, I have not given an extensive biography of Oresme, nor bibliography of his works. These are available elsewhere, as indicated in the notes at the appropriate places in the introduction, and need not be repeated in detail here.

Finally I would like to thank the librarians at St. John's College, Oxford, for sending me photographs of the appropriate folios of ms. 164; Dr. Maureen Boulton, who has read the text and made numerous comments on the Middle French, and more or less kept me from going astray; and the anonymous reader who made a number of suggestions that have made this edition more useful to its readers.

from Proquest or a similar source.

His doctoral dissertation was "Jehan de Lanson, Chanson de Geste of the Thirteenth Century,edited after the manuscripts of Paris and Bern with introduction, variants, notes, table of proper names, and glossary," University of North Carolina at Chapel Hill, 1959; published: Chapel Hill: University of North Carolina Press, 1965.

1
NICOLE ORESME

The Middle French *Traitié de l'espere* was written by a well-known medieval cleric, Nicole Oresme, bishop of Lisieux (ca. 1320–1382). A number of detailed, and still relevant biographies are available,[1] so that only an outline of his career will be given here.

Born around 1320, probably in northern France (possible around Caen), he seems first to appear in 1342 as a regent master of arts,[2] and then in 1348 in a scholarship list in theology at the College of Navarre, Paris. He completed his theological mastership in 1355 or 1356 and became grand master of the College of Navarre in 1356. About this time he became friends with the dauphin of France (the future King Charles V). His name appears in a few court documents of the period while the dauphin was acting as regent for his father. In 1362 he became a canon of Rouen, then a canon of the Sainte-Chapelle, Paris, and then (1364) dean of the cathedral of Rouen. In 1377 he was appointed bishop of Lisieux. He died in 1382.

Oresme's writings span a broad range of subjects. Marshall Clagett describes him "as a subtle Schoolman disputing the fashionable problems of the day, a vigorous opponent of astrology, a dyanmic preacher and theologian, and a skillful translator of Latin into French"[3] including a number of works of Aristotle.

His writings are comprised of a number of works on mathematics and astronomy; natural philosophy (including several commentaries on the works of Aristotle – *Physica*, *De caelo*, *De generatione et corruptione*, *De anima*, *Meteorum*, *Parva naturalia*); moral and social philosophy (e.g., the *Ethica* and the *Politica* of Aristotle); economics and money theory; astrology (more correctly, against astrology); and theology and sermons.[4]

1 See: Marshall Clagett, "Oresme, Nicole," *Dictionary of Scientific Biography*, ed. Charles C. Gillispie (New York: Scribner's, 1974), 10: 223-230. Benoît Patar, *Nicolai Oresme, Expositio et Quaestiones in Aristotelis De anima* (Louvain-la-Neuve: Éditions de l'Institut Supérieur de Philosophie, Louvain-Paris: Éditions Peeters, 1995), pp. 12*-14*. The most detail is given in Edward Grant, *Nicole Oresme, De proportionibus proportionum and Ad pauca respicientes* (Madison: Unversity of Wisconsin Press, 1966), pp. 3-10. There is also a very useful biography in Nicole Oresme, *Le Livre du ciel et du monde*, ed. Albert D. Menut and Alexander J. Denomy (Madison: University of Wisconsin Press, 1968), pp. 3-9. One more recent piece of information is reported by William J. Courtenay, "The Early Career of Nicole Oresme," *Isis* 91 (2000), 542-548.

2 Courtenay, pp. 542, 544.

3 Clagett, *DSB*, p. 223.

4 The most complete list of Oresme's works can be found in Daniel A. Di Liscia and Aurora Panzica, "The Writings of Nicole Oresme: A Systematic Inventory." *Traditio* 77 (2022), 235-375.

Among the cosmological/astronomical works, those most relevant to the *Traitié de l'espere* would be:[5]

Questiones super De celo (1348-1350)
Questiones super De spera (1348-1350)
Questiones super De spera (Erfurt ms)[6]
Ad pauca respicientes [or possibly *De motibus sperarum*] (1351-1360)
De commensurabilitate vel incommensurabilitate motuum celi (1350-1360)
Le livre de divinacions (ca. 1356/1366)
Le livre du ciel et du monde (1377)

Within this set the *Traitié de l'espere* might be dated as probably between 1356 and 1377.

During most of his life Oresme enjoyed a strong relationship with Charles V both as dauphin and regent for his father Jean II (while held captive in England, 1356-1360), and as king (1364 onwards). The association seems to have been as early as 1359 and continued through to the death of the king in 1380. As chaplain and counsellor Oresme undertook a number of state missions for the king, and was suitably rewarded by the king for his work.

Charles was interested in educating himself and his entourage, his council and others, in the works of Aristotle and others, deeming that difficult topics would be better understood in French rather than Latin.[7] Thus he commissioned from Oresme translations of Aristotle's *Ethica Nichomachea*, *Politica*, and *De caelo*, as well as the *Oeconomica* ascribed to Aristotle. (Oresme also translated other works into French, and Charles also commissioned translations by other scholars.)[8] As part of this program of enlightening the upper echelon of French society, Oresme wrote the *Traitié de l'espere* on astronomy / cosmology which was later supplemented by his *Livre du ciel and du monde*.

5 Dating is from Di Liscia/Panzica and Benoît Patar, pp. 28*-29*

6 Alain Boureau, who has recently edited these questions, argues that they are indeed by Oresme and possibly written about 1340, early in Oresme's career. See Alain Boureau, ed., *Nicole Oresme, Deux suites de questions sur la sphère ...*, Astronomie, dioptrique et cinétique tome x (Paris: Les Belles Lettres, 2024), pp. 118-121.

7 Menut and Denomy, p. 3.

8 Menut and Denomy, p. 6.

2
THE *TRAITIÉ DE L'ESPERE*

The Purpose of the Traitié

The *Traitié de l'espere* was written by Oresme as part of Charles V's program of educating himself and his councillors. Oresme himself writes in his prologue:

> The shape and disposition of the world, the number and order of the elements, and the movements of celestial bodies are appropriate for the knowledge of every man who is of free status and of excellent intelligence, and this is a good thing and good to know and delectable, profitable and honest, and with this it is necessary for knowing philosophy, and especially astronomy/astrology.[1]
>
> (*lines 3-7*)

and in his final chapter:

> My intent is not to go further in these subtleties nor to treat astrology [i.e., astronomy] in French, but only to explain roughly the disposition of the sphere of the world in general and of this that which is suitable knowledge for every man and especially for the prince of excellent intelligence, so that he does not leave undone anything of this which appertains to his office or status, and to the governance of public affairs.
>
> (*lines 1564-1571*)

With this in mind, Oresme has created a very basic (but not unsophisticated) text, an introduction to astronomy and cosmology, focussing on the physical universe rather than the mathematical.

The Contents of the Traitié

Oresme begins by telling us that the universe, the earth, and likewise the moon are spherical. At the very centre is the earth – the element earth – covered where it

1 Like others in the Middle Ages, Oresme often uses the word "astrology" and "astronomy" interchangeably. Both terms generally mean the study of the stars, of their motions and of their reciprocal relations (often with the aim of determining their influence on human destiny, weather, etc.). Similarly with "astrologer" and "astronomer".

is lowest by water, and where the earth protrudes above the water is the area in which animals and humans live. The earth and water is enclosed in a sphere of air, above which is a layer or sphere of pure fire. Finally is the sphere of the heavens in which we find the stars (and planets) and which revolves around the earth from east to west. (*Chap. 1*)

The heavens are transparent, and it and the stars are immutable. It is pure (unlike the spheres of the four elements) but can affect the lower spheres by its light, its motion, and its "influences". (*Chap. 2*)

The heavens, however, are divided into seven sub-spheres, one for each of the planets (including the sun and the moon) which he lists and beyond which is an eighth sphere for the fixed stars and a ninth sphere for God and his entourage. Oresme also describes how the inner concave surface of each sphere of a planet encloses the outer convex surface of the sphere of the planet below it (or, as for the moon, the spheres of the four elements below it). (*Chap. 3-4*)

The daily motion of the heavens is as if it revolves around an imaginary axis through the earth below it, an axis which pierces the earth at the poles of the earth (north and south, arctic and antarctic). Halfway between the two poles and around the earth is the equinoctial circle or equator. (*Chap. 5*)

The planets are carried around the earth by the daily motion of the heavens, but they also have a second motion through the heavens. The period of these second motions varies, from 27 days 8 hours for the moon[2] to about 30 years for Saturn. (*Chap. 6*)

While the daily motion of planets (including the sun) is measured along the equinoctial circle, the secondary motion is at an angle to the earth's equinoctial circle, within what is known as the zodiac. The zodiac (more properly its centre line – the ecliptic) intersects the equinoctial circle at two opposite points, and the sphere of the zodiac has its own set of "poles." There are four main points along the zodiac (or ecliptic) – where it intersects with the equinoctial circle (the equinoxes) and where it turns back south or north (the solstices). Each of these four arcs is divided into three sub-arcs known as "signs" which Oresme lists and relates to their position vis-à-vis the four points on the ecliptic. Each sign is divided into 30 parts or degrees (for a total of 360 degrees), and each degree into 60 minutes, each minute into 60 seconds, and so on. (*Chap. 7 and 8*)

These divisions are re-iterated in Chap. 9, with a comment that the sexagesimal divisions are arbitrary and not "by nature", set out by humans for ease of measurement and calculation.

The ecliptic has a length of 360°, and a width or latitude of 12°; the sun travels along a line in the middle of this band, called the "ecliptic." The width of the band allows for the variation in latitude of the orbits of the other planets. (*Chap. 10*)

2 This is the siderial period (27 days, 7 hours, 43 minutes) measured against the fixed stars.

The concept of a "sign" can be understood in three ways. First it is a section of the "surface" of the zodiac, 30° long and 12° wide. Second it can designate everything within this area of the zodiac, regardless of its distance from the earth. Third, it can designate a spherical wedge (or ungula) of the whole sphere of the universe, the sphere being divided into 12 parts from pole to pole of the earth, and everything within this wedge regardless of its distance from the earth or from the ecliptic. The first meaning is the most proper. (*Chap. 11*)

Two colures also divide the sphere, the first through the poles of the earth and the poles of the ecliptic as well as the two solstices, and the second through the poles of the earth and the two equinoxes. The first colure passes through the beginning of Cancer and the beginning of Capricorn which are a little more than 23° from the equatorial circle (as is the distance between a pole of the earth and the nearby pole of the ecliptic). (*Chap. 12*)

The meridian circle passes through the poles of the earth and the point directly overhead the observer (called his "zenith") and the sun is found along the meridian at his mid-day. When the sun is on the opposite side of the earth, it would be midnight. If the observer moves east or west, his meridian moves with him If he stands still or moves north or south, the meridian remains the same. (*Chap. 13*)

The other circle which divides the heavens is the "horizon", the circle above which we can see the heavens and below which we cannot. The centre of a person's horizon is the person's zenith, and however that person moves (north/south or east/west) his zenith and therefore his horizon moves with him. There are two types of horizons, a right horizon and an oblique horizon. A right horizon is viewed by anyone on the equator whose zenith will be on the equatorial circle. Someone not on the equator has an oblique horizon, or a horizon at an angle; within this horizon one or other of the poles is visible. (*Chap. 14 and 15*)

The elevation of the pole is its angular distance above the horizon, which is also the angular distance between the observer and the equinoctial circle (i.e., the latitude). The solstice points are about 23° from the equinoctial circle, so at the "summer" solstice the sun is at an angle 23° less than the latitude of the observer and at the "winter" solstice it is at an angle 23° more than the latitude of the observer.[3] (*Chap. 16*)

There are also four "lesser" circles besides those previously mentioned, lesser in the sense that they are shorter in circumference but still divided into 360 degrees. Two of these are the tropics, through the solstices, which are the limits of the sun's north/south annual motion. (Oresme calls them the "summer tropic", i.e., Tropic of Cancer, and the "winter tropic", i.e., the Tropic of Capricorn.) As well there are the Arctic and Antarctic Circles near the poles of the earth. These four circles are parallel to the Equator. As well the distance of the Arctic and the Antarctic Circles

3 For Oresme, "summer" and "winter" always refer to the seasons in the northern hemisphere.

to their respective poles is the same as each of the tropics is from the Equator, i.e., around 23 degrees. (*Chap. 17)*

The natural day is the time the sun takes to make a complete circuit of the earth from, say, noon one day to noon the next. The artificial day is the time between sunrise and sunset on any particular day. The motion of the latter (sunrise to sunset or above the horizon) is the "arc of the day" and from sunset to sunrise (below the horizon) is the "arc of the night". These arcs vary, according to the distance of the sun from the equinoctial circle – sometimes the arc of the day is greater than the arc of the night and sometimes vice versa, depending on the season. In any event the relative length changes each day; it also varies with the latitude of the observer. Only at the equinoxes are the two arcs equal. (For someone at the equator, the two are always equal.) (*Chap. 18*)

Because of the angle of the ecliptic and the angle of the horizon, the signs in a given quarter of the zodiac rise above the horizon and set at different times vis-à-vis the equinoctial circle, that is, one twelfth of the zodiac can rise before one twelfth of the equinoctial circle, and vice versa. A change can also occur in different seasons. Since half the zodiac rises and sets within the arc of the day, when the days are shorter they rise and set more quickly; when the days are longer they rise and set more slowly. (*Chap. 19 and 20*)

Also the risings and settings of the signs are displayed differently at different latitudes – e.g., at the Arctic Circle, between the Arctic Circle and the North Pole, at the North Pole, at the Equator. (*Chap. 21*)

The orbit of the sun is not centred on the earth but is eccentric with an apogee and a perigee. But Oresme says that the nature of the sun's eccentricity is beyond the scope of this text. (*Chap. 22*)

Furthermore, because of the eccentricity of the sun's orbit, there is a difference between the two "halves" of the year, that the period between the March solstice and the September solstice is longer than the period between the September solstice and the March solstice.[4] (*Chap. 23*)

Because of the obliquities of various circles, and the eccentricity of the sun, as well as the proper motion of the sun along the ecliptic, the natural (24-hour) days vary slightly, being slightly longer in September than in March. (*Chap. 24 and 25*)

Oresme next looks at the size or circumference of the earth (as well as the diameter). He uses Sacrobosco's measure of 87.5 miles per degree, taken from Eratosthenes, even though he will point out in Chap. 34 that Sacrobosco changes his measure to that of al-Farghānī (56.66 miles).[5] Oresme also lays out his table of measures for leagues, miles, stadia, strides, feet, palms and digits. (*Chap. 26*)

---

4 See the addenda to Chap. 23, after line 683.

5 These are not modern "English" or "American" miles, but medieval French or even ancient "miles".

Oresme next focuses on the habitable regions of the earth in a number of chapters. First he looks at the how the sun is viewed in various regions from equator to the pole, in terms of the length of the day and variations because of the seasons. He then examines the five regions of the earth (North Pole to Arctic Circle, Arctic Circle to Tropic of Cancer, between the two tropics, Tropic of Cancer to Antarctic Circle, Antarctic Circle to the South Pole). Subsequently he discusses how one should determine which pole is "on top" of the world, and which is "at the bottom". (*Chap. 27 and 28*)

Oresme then discusses the properties of the third region (between the two tropics) – length of days, seasons, excessive heat; and the properties of the fourth region (Tropic of Capricorn to the Antarctic Circle) and whether there are inhabitants in the Antipodes – arguments concerning theology, climate, and whether that section is in fact covered with water. (*Chap. 29 and 30*)

Following this he presents the divisions of the earth according to ancient writers (*Chap. 31*) and according to astrologers/astrologers – first of all Ptolemy who divides the earth into 27 regions (by lines of latitude from the equator north) according to the lengths of the longest day in each band. More astronomers, e.g. al-Farghānī, divide the habitable world into seven bands, or "climes" with names derived from the ancient world, e.g., "through Alexandria", "through Rhodes", etc.,[6] as well as the association of the climes with the planets. (*Chap. 31 and 32*)

Next Oresme lists the length of each clime, or the circumference of each clime, in degrees, in miles and in leagues. Here Oresme points out that there are two values for a degree, 87.5 miles and 56.66 miles (from al-Farghānī). He also lists the width of each clime, again pointing out the two different values (both of which are used in Sacrobosco's *De spera*). (*Chap. 33 and 34*)

For each clime, Oresme gives the length of the longest day at the beginning and at the end (Sacrobosco, gives the longest day at the mid-point of the clime). The climes are divided by half-hour differences in the longest day, and Oresme criticises Sacrobosco for saying (essentially) that a 15-minute difference would be half-way across the clime, whereas the distance across of the second 15-minute division would in fact be less than across the first 15-minute division of the clime, and similarly for any other division of a clime. (*Chap. 35*)

Finally Oresme gives the elevation of the pole for each clime (beginning, middle, end) which is in fact the latitude of the beginning, middle and end of each clime. (*Chap. 36*)

Next Oresme treats of the area, or the amount of the Earth that is habitable, and then of those areas which might be habitable outside of the seven climes (north and south, and also east and west). (*Chap. 37 and 38)*

Then Oresme turns to a phenomena which occurs if one were to travel around

---

6 See the appendix on "Climes."

the world either westward or eastward. Because the daily movement of the traveller also moves his meridian and his "noon", the traveller ends up counting one less "day"[7] (when moving westward) and one more "day" (when moving eastward) than would be counted if he had stayed home.[8] (*Chap. 39*)

Finally Oresme examines the limitations of habitable lands – what determines if a land is habitable (*Chap. 40*); the nature of creatures at the extremes of the habitable world (*Chap. 41*); differences in the nature of humans according to the latitude of their habitations (*Chap. 42*) and according to the longitude of their habitations (*Chap. 43*)

Oresme then moves on to the nature of the phases of the moon, analysing its relationship with the sun. (*Chap. 44 and 45*)

Then he treats eclipses in general (*Chap. 46*), of the sun (*Chap. 47*), of the moon (*Chap. 48*), of the colours of the moon during an eclipse, and eclipses of stars (*Chap. 49*).

Chapter 50 closes the text – an explanation, again, of why Oresme wrote the *Traitié*, and why he did not include other matters in it. At the very end he appends an alphabetical table of terms used in the treatise, and where a reader might find the first definition of those terms (chapter and part of chapter).

The **Traitié** *and the* **De spera** *of Sacrobosco*

The *De spera* of Sacrobosco, written about 100 years earlier, is much more well known than Oresme's *Traitié*, and was so even in the middle ages. There has been some early confusion between the two texts, and at least one early publisher considered the *Traitié* to be a translation of the Latin *De spera*.[9] However the *Traitié* is significantly different from Sacrobosco's work, even though they cover the same material.

Sacrobosco's first chapter[10] on the spherical universe and its parts is covered in much more detail by Oresme in his first six chapters (where it is better organized). Sacrobosco's second chapter deals with the range of circles which divide the heavens (equinoctial circle, ecliptic, horizon, tropics etc.) which Oresme expounds in chapters 7 to 17, again in much more detail, although somewhat in the same order.

The third chapter of the *De spera* treats the rising and setting of the stars, etc.; length of the days and nights; movement of the sun; and the inhabitants at various

7 By "day" he does not mean the standard 24-hour day, but the number of sun-rises or sun-sets or "noons" that the traveller experiences on his trip.

8 This is the phenomenon which is the basis of Jules Verne's 1878 novel, *Around the World in Eighty Days* (*Le Tour du monde en quatre-vingts jours*).

9 Johannes de Sacro Bosco, *Le traicté de la sphere*, translaté de latin en françois par maistre Nicole Oresme (Paris: Chez Simon du Bois pour Regnault Chaudière, [1528]). (USTC 8403)

10 These are the chapters as set out in Thorndike's edition: Lynn Thorndike, *The Sphere of Sacrobosco and Its Commentators* (Chicago: University of Chicago Press, 1949).

points in the northern hemisphere, as well as the seven climes and beyond. These are covered in much more detail (especially an analysis of the "climes") by Oresme at various points from chapter 19 through 36[11] – but in these chapters he also deals with other matters not really touched on by Sacrobosco.

The last chapter of the *De spera* treats the motion of the sun and of the moon and other planets, as well as eclipses. These too are covered by Oresme in more details near the end of his text, but he also covers in detail a number of other related topics such as the limits of inhabitable lands (Chap. 37, 38, 40-43) and travelling around the world (Chap. 39). An example of the difference in treatment, Sacrobosco devotes 8 lines to the lunar nodes while Oresme spends three times that space explaining this aspect in more detail and relating them to solar and lunar eclipses.

In terms of the two texts, it is as if we have two introductory texts on astronomy, but with Oresme's text dealing with the subject more comprehensively and in a more coherent order. Oresme's *Traitié* in no way is a French version of the *Sphere*, let alone a translation.

The **Traitié** *and the* **Livre du ciel et du monde**

Oresme composed his *Traitié de l'espere* several years before his *Livre du ciel et du monde*. The latter text purports to be a translation of Aristotle's *De caelo* (using the *nova translatio* of Robert of Lincoln and William of Moerbeke), but while faithful to the meaning of Aristotle, the text is more a paraphrase (and commentary), and it depends heavily as well on Averroës Middle Commentary.[12]

The *Livre* is a combination of a translation and Oresme's own commentary, but of course it is governed by the contents of the Latin original. In some places, therefore, Oresme felt he needed to refer to an explanation or definition found in his earlier *Traitié*, and he refers to the earlier text many times ("It is explained in our *Treatise on the De spera* ...").[13] Thus in a sense the *Traitié* became a sort of prologue or introduction to the *Livre*, and as such they are found together in some of the manuscripts.[14] In fact Oresme even suggests that the two texts be bound together.[15]

11 Oresme correctly points out errors by Sacrobosco in the sections dealing with the values for the length of a degree (Chap. 34). He also criticises Sacrobosco's analysis of the length of the longest day in each clime (see Chap. 35, notes to lines 1102 and 1105); see below, appendix on "Climes."

12 *Le Livre,* ed. Menut and Denomy, pp. 10-11.

13 E.g., *Le Livre,* ed. Menut and Denomy, pp. 353, 487, 563 etc.

14 Four of the five manuscripts of *Le Livre* contain the *Traitié* as well. See below, my mss A, F, G, and L.

15 "... I have finished the First and Second Books of the *De caelo et mundo*, for the better understanding of which it is expedient to read also the *Treatise on the Sphere* in French, which I have previously mentioned. And it would be good if it could be put in one volume along with these two books, and, it seems to me, this would make a noble and very excellent book of natural philosophy." *Le Livre,* ed. Menut and Denomy, p. 181.

The Questiones

Among Oresme's other astronomical works are several sets of questions. The first is the *Questiones super De spera*[16] containing 13 questions which Gerrett Droppers characterizes as "a series of original studies or *questiones* on particular problems, implicit or explicit, in Sacrobosco (or perhaps some other elementary treatise) which are in all likelihood too sophisticated and problematical for an elementary class."[17] The topics are the nature of lines and points (generally in a mathematical sense) (qq. 1 and 2); the Aristotelian doctrine of place and of heaviness and lightness (qq. 3-6); and a dynamic and kinematic analysis of terrestrial and celestial motion (qq. 7-13). These topics are completely unrelated to those found in the *Traitié de l'espere.*[18]

The second set of 45 questions (in four books), entitled *Questiones super De celo*,[19] fits into the tradition of questions and commentaries on Aristotle written by medieval schoolmen. Like other sets of questions by Jean Buridan and Albert of Saxony, Oresme explores philosophical issues arising from Aristotle's text such as the plurality of worlds, space outside the universe, notions of infinite divisibility and the infinite itself (Book I). In Book II he deals with celestial bodies as well as the cause of acceleration and the possibility of the earth's diurnal rotation while in Book III he deals with the elements. Book IV treats hydrodynamical issues arising from the concepts of relative and absolute heavy and light. Again these philosophical discussions are distant from the cosmographical focus of the *Traitié de l'espere.*[20]

There is also a third set of 17 questions on the sphere found only in one manuscript (Erfurt, Universitäts- und Forschungsbibliothek Erfurt/Gotha, ms Amplon. Q° 299) which has recently been transcribed and published with a French translation.[21] The attribution to Oresme has been debated since its discovery in 1961[22] but if by Oresme, it is an early work, composed possibly about 1340 when Oresme was in his late teens or early 20s. The text follows the standard scholastic form of this genre: proposition, statements supporting and contradicting the proposition, a discussion of each of these statements, and the author's (Oresme's) conclusion. The questions focus on specific points: the mathematical and physical nature of a sphere, the celestial sphere or spheres (its motion, the number of spheres, are the spheres truly

---

16 Edition: Garrett Droppers, "The Questiones De sphera of Nicole Qresme," PhD. thesis, University of Wisconsin, 1966.

17 Droppers, p. 5.

18 Ibid.

19 Edition: Claudia Kren, "The Questiones super *De celo* of Nicole Oresme," PhD. thesis, University of Wisconsin, 1965.

20 Kren, pp. xv-xvt.

21 Alain Boureau, ed., *Nicole Oresme, Deux suites de questions sur la sphère* ..., Astronomie, dioptrique et cinétique tome x (Paris: Les Belles Lettres, 2024), pp. 137-371.

22 Di Liscia/Panzica, pp. 329-330: "Writings whose Oresmian Attribution has not been established".

spherical), the earth (is it spherical, do its poles move). There are also questions about a variety of celestial circles – the equatorial circle, the tropics, the colures, the zodiac, its inclination and how it is divided; and three questions on whether certain zones of the earth are habitable. Finally there are three questions about the eccentric motion of the sun, the moon's luminosity, and eclipses. The thrust of these questions is generally about proving the accepted (often mathematical) nature of these topics. In the *Traitie de l'espere* these points are for the most part assumed, and the conclusions stated as facts without proof. While the text can be seen as background to the *Traitie*, it treats these topics more narrowly although much more extensively.

The Middle French[23]

Oresme's text of the *Traitié* was written in what is now classified as Middle French, whose vocabulary is very close to modern French. In Middle French, noun declensions have been eliminated, and plurals for the most part are created by adding an *s* (or a *z*). Sentence structure or word order becomes more important for communicating meaning, much like in modern French.

The main difference between the two is in the orthography, since spelling was not standardized until the seventeenth century. The differences for any one word reflects the variety of ways in which a spoken word can be represented by letters: *s/c, i/y, f/v, n/ng* etc. As well, letters which are not normally pronounced (final *s*, final feminine *e*, or in words like *tel/telle*, i.e., double consonants) were often not written out. These variations sometimes result from local dialect or accent, which changes the pronunciation of a word and therefore its spelling.

It is now common practice in modern editions of Middle French texts not to list these orthographic or dialect variations as found in different manuscripts but to concentrate on the substitution of words, or additions or deletions which might affect the meaning. As well, since there were no accents in Middle French, the acute accent only is used to distinguish between different meanings of the same word or for a tonic final *e* (aprés, eclipsé, etc.).

One other element worth noting is that in this period new words were being introduced to French (often a variation on the word in Latin), particularly for technical terms or ideas. Oresme himself created many of these, out of necessity, in his various works, and a list of those which appear to have been created while writing the *Traitié* has been added below in an appendix.

23 For McCarthy's analysis of the dialects of seven of the manuscripts, see below, "3. The Manuscripts and Editions," note 1.

3
THE MANUSCRIPTS AND EDITIONS

The Manuscripts[1]

The *Traitié de l'espere* survives in 13 manuscripts:

A Bern, Burgerbibliothek (*olim* Bibliotheca Bongarsiana), ms 310, ff. 1ra-27r (late 14th century).[2] Begins Chap. 12, line 354.
Also contains (ff. 28ra-152vb) Oresme, *Le Livre du ciel et du monde*. [McCarthy ms E.]

B Bordeaux, Bibliothèque municipale, ms. 531, ff. 90r-127r (fol. 127r: "anno domini 1448").

C Firenze, Biblioteca Medicea Laurenziana, ms. Ashburnham 1604, ff. 1r-39v (end 14th century).[3]
Also a very well written manuscript, and close to the base text in ms E.

D Leiden, Bibliotheek der Rijksuniversiteit, ms. Vossius gall. folio 10, ff. 1r-31v (15th century).[4]

E Oxford, St. John's College, ms. 164, ff. 1r-32r (ca A.D. 1365-1377).[5]
This is an elaborately decorated text, in a textura rotunda hand, well above

---

1 McCarthy in her thesis (pp. 37-72) examined the dialects of (my) mss AFGHKLM ("vowels", "consonants", and for ms H also "morphology and syntax"). Her hand-written note in the copy in the PIMS Library (p. 72) reads "From the foregoing analysis, it is obvious that the manuscripts are written in one of the northern dialects, more specifically, north-western."
Readers interested in these aspects should consult the thesis.

2 J. R. Sinner, *Catalogus codicum mss. bibliothecae Bernensis* (Bern, 1760), 1: 284-285; Hermann Hagen, *Catalogus codicum Bernensium* (Bern, 1874), pp. 322-323; Nicole Oresme, *Le Livre*, ed. Menut and Denomy, p. 34

3 Gilbert Ouy, "Jean Lebégue (1368-1457), auteur, copiste et bibliophile," in Godfried Croenen and Peter Ainsworth, eds., *Patrons, Authors and Workshops* (Louvain: Peters, 2006), pp. 161-162.

4 J. P. Gumbert, "Medieval Manuscripts in French in the Leiden University Library: A Handlist," in Peter Rolfe Monks and D.D.R. Owen, eds., *Medieval Codicology, Iconography, and Translation*, (Leiden: Brill, 1994), p. 38.

5 Ralph Hanna and Jeremy Griffiths, *A Descriptive Catalogue of the Western Medieval Manuscripts of St John's College Oxford* (Oxford: University Press, 2002), pp. 227-228. See also Edgar Laird, "Astrology in the Court of Charles V of France, as Reflected in Oxford, St John's College, ms 164," *Manuscripta* 34 (1990), 167-176.

the standards of the other manuscripts. It is possibly the presentation copy to Charles V. My base manuscript; also McGovern's base ms; not in McCarthy or Myers. NOTE: the scribe writes "u" and "n" virtually identically; I have generally transcribed them to fit the normal spelling of the word.

F Paris, Bibliothèque nationale de France, ms. fr. 565 (*olim* Anc. 7065), ff. 1r-22v (early 15th century).[6]

Also contains (ff. 23ra-171vb) Oresme, *Le Livre du ciel et du monde.* [McCarthy ms B; Myers ms B.]

Abbreviations here mimic the abbreviations in medieval Latin, e.g., "ē" for "est" and "p" for "per/par/per-/par-". The ms also uses a line over a vowel to indicate a following "m" or "n", e.g. "mōde" for "monde". "Et" is abbreviated using medieval Latin forms.

G Paris, Bibliothèque nationale de France, ms. fr. 1083 (*olim* Colbert 3230; *olim.* Anc. 7350(2)), ff. 126ra-145rv (early 15th century).[7]

Also contains (ff. 1ra-125rb) Oresme, *Le Livre du ciel et du monde.* [McCarthy ms G; Myers ms C.]

H Paris, Bibliothèque nationale de France, ms. fr. 1350 (*olim* Colbert 4270; *olim* Anc. 7483(6)), ff. 1ra-38v (end of 14th century).[8]

[McCarthy ms A – her base text; Myers ms A – his base text.]

K Paris, Bibliothèque nationale de France, ms. fr. 2240 (*olim* de la Mare 222; *olim* Anc. 8014(2)), ff. 61r-96r (15th century).[9]

[McCarthy ms F.]

L Paris, Bibliothèque nationale de France, ms. fr. 24278 (*olim* St. Victor 328), ff. 140r-146r (14th century).[10] Begins Chap. 37, line 1150.

Also contains (ff. 1r-138v) Oresme, *Le Livre du ciel et du monde.* [McCarthy ms D.]

6 See *Le Livre*, ed. Menut and Denomy, p. 33 note 4.

7 See *Le Livre*, ed. Menut and Denomy, p. 34.

8 François Avril (card file at the Bibliothèque nationale) dates this as 14th century, as does the website "Jonas" (Institut de Recherche et d'Histoire des Textes, Répertoire des textes et des manuscrits médiévaux d'oc et d'oïl). The Bibliothèque itself (Département des Manuscrits) describes it on-line as 15th century, as does Coopland, quoting a French source but not identifying it. (G. W. Coopland, *Nicole Oresme and the Astrologers* [Cambridge, MA: Harvard University Press, 1952], p. 49.)

9 François Avril (card file at the Bibliothèque nationale); Bibliothèque nationale de France (Département des Manuscrits) – on-line description.

10 See *Le Livre*, ed. Menut and Denomy, p. 35.

M Paris, Bibliothèque nationale de France, ms. nouv. acq. fr. 1052, ff. 1r-38r (A.D. 1484-1485).[11]

[McCarthy ms C.]

This copy includes a large number of longer, unique variations. As well, some of the chapters (or parts of chapters) are out of order – as if the scribe dropped his exemplar and re-assembled it haphazardly. E.g., lines 585-629 (Chap. 21) have been moved from f. 15r to ff. 13r-14r, after line 537 (Chap. 19). E.g., f. 18v – the end of Chap. 26 (to line 774); ff. 18v-19r – Chap. 28 (lines 825-852); ff. 19r-19v – Chap. 29, the first part (lines 858-872); ff. 19v-20r – Chap. 27 (lines 776-814); ff. 20r-20v – Chap. 28 (lines 815-823); ff. 20v-21r – Chap. 29, end (lines 873-875); ff. 21r and following – Chap. 30 etc.

N Paris, Bibliothèque nationale de France, ms. nouv. acq. fr. 10045, ff. 1r-39v (14th century).[12]

P Vatican, BAV, Reg. lat. 1337, ff. 29r-44v (15th century).[13]

Close to the base text in ms E. Generally replaces roman numerals with arabic numerals (or replaces roman numerals with words).

Ghost reference

Paris, Bibliothèque nationale de France, ms. fr. 7487.

This manuscript contains documents covering the years 1302 to 1305.

In 1857 Francis Meunier[14] listed a copy of Oresme's text in "anc. f. français, ms. n° ... 7487" which began a thread of misinformation. Menut and Denomy[15] converted this to simply "fr. 7487" (dropping the crucial "ancien fonds" information) which was repeated in Patar's list.[16] Ms "fr. 7487" also appears in several places on-line.

Ms ancien 7487 is now ms fr. 1357. Meunier noted that this manuscript contained only extracts, made under Charles VIII, by one "Symon de Phares". Ms fr. 1357 is, in fact, a unique copy of Simon de Phares' *Re-*

---

11 Françoise Fery-Hue, "Le mathématicien Nicolas Chuquet et la planche à régler," *Pecia* 13 (2010) [= *Du scriptorium à l'atelier*], p. 335.

12 Henri Omont, *Bibliothèque nationale. Catalogue général des manuscrits français. Nouvelles acquisitions françaises, 4: nos 10001-11353 et 20001-22811* (Paris, 1918), p. 10.

13 Ernest Langlois, "Notices des manuscrits français et provençaux de Rome antérieurs au xvie siècle," *Notices et Extraits des Manuscrits de la Bibliotheqe Nationale et Autres Bibliothèques* 33 part 2 (1889), 131: "diverses écritures du xve siècle."

14 Francis Meunier, *Essai sur la vie et les ouvrages de Nicole Oresme* (Paris, 1857), p. 58.

15 A. D. Menut and A. J. Denomy, "Maistre Nicole Oresme, Le Livre du ciel et du monde," *Mediaeval Studies* 5 (1943), 247. They note that this copy ends with chapter 17.

16 Benoît Patar, p. 19*

cueil des plus celebres astrologues (ff. 1r-166r)[17] but this treatise does not contain any text of Oresme, a noted critic of astrology, nor is any such text appended to Simon's treatise in the manuscript.

Editions

The text seems to have been printed once in the sixteenth century:

Johannes de Sacro Bosco, *Le traicté de la sphere*, translaté de latin en françois par maistre Nicole Oresme (Paris: Chez Simon du Bois pour Regnault Chaudière, [1528])

USTC 8403

While ascribed to Sacrobosco, it is actually the text by Oresme and not a translation. The fact that the date is unclear seems to have led some people like Francis Meunier[18] and Pierre Duhem[19] to assume that it was printed earlier or that there was an earlier edition. However, Simon du Bois / Simon Dubois carried on business in Paris only between 1525 and 1529.[20] Brigitte Moreau dates this publication to "c. 1528, d'après le matériel et les bois"....[21]

The text was edited (but not published) twice in the 1940s and once in 1974:

John Vernon Meyers, ed., "Maistre Nicole Oresme, Le Traité de la Sphère," Master of Arts thesis, Department of Romance Languages, Syracuse University, 1940.[22]

Lillian McCarthy, ed., "Maistre Nicole Oresme, Traitié de l'Espere," Doctor of Philosophy thesis, University of Toronto, 1943.[23]

Terrence Joseph McGovern, ed., "Traité de l'Espere and Minor Astrological

---

17 The text has been edited: Jean-Patrice Boudet, *Le* Recueil des plus celebres astrologues *de Simon de Phares*, 2 vols. (Paris: Champion, 1997-1999).

18 Francis Meunier, *Essai sur la vie et les ouvrages de Nicole Oresme* (Paris, 1857), p. 58.

19 Pierre Duhem, *Le système du monde* (Paris, 1916), 4: 160 n 2.

20 D.J. Shaw, "New Dates in the Career of Simon Du Bois, Reformation Printer, Paris 1523-1529," *Yale University Library Gazette* 67 (1992-1993), 32-36.

21 Brigitte Moreau, *Inventaire chronologique des éditions parisiennes du xvie siècle* (Abbeville: F. Paillart,1985), vol. 3, entry 1574.

22 Copies currently in the University of Syracuse Library, and the library of the Pontifical Institute of Mediaeval Studies in Toronto.

23 Copies currently in the University of Toronto Library (2 copies), the library of the Pontifical Institute of Mediaeval Studies in Toronto, and Library and Archives Canada (microfilm copy).

Di Liscia/Panzica (p. 292) erroneously state that McCarthy included an English translation; she did not.

Works: A Critical Edition," Doctor of Philosophy thesis, The Catholic University of America, 1974.[24]

Translations

None of the editions mentioned above have been translated into a modern language (although the Middle French is relatively easy to read on its own). However, there is a curious translation made into Latin in the fifteenth century. It survives in a unique copy:

New Haven, Yale University Library, Beinecke ms 335, ff. 15v-37v

I have taken samplings from this Latin text and found that it does closely match the original Middle French version. The drawings are particularly well done.

The Base Manuscript

I have chosen ms E (Oxford, St. John's College, ms. 164) as the base manuscript for a number of reasons. First of all it has been dated as one of the earliest manuscripts – if not the earliest – 1367 X 1377.[25] Three other manuscripts are dated as "late 14th century" (ms A) or "end of 14th century (ms C – a text close to E; ms H). Two are dated "14th century (ms L and ms N). Mss A and L must be later since they both also contains a copy of Oresme's *Livre du ciel et du monde*, written after the *Traitié*.[26] The other manuscripts – BDFGKMP – all date from the 15th century.

Second, it is generally agreed that ms E is most probably a presentation copy to King Charles V (reigned 1364-1380), written in a textura rotunda hand.[27] Following Oresme's treatise are others considered to be of interest to Charles and composed in or translated into French for his benefit.[28] On fol. 1r there is an image of Charles with a book on his lap facing a table with other books and an armillary sphere. As well on fol. 33r there is a second image of the king receiving a book from

---

24 This is not, in fact, a critical edition but only a transcription of the text in Oxford, St. John's College, ms. 164, as well as of other texts in that manuscript. Copies are available on line. Since this is also a transcription of my base text, I have compared my readings with those of McGovern's, and with the manuscript. I have discovered a number of faulty readings by McGovern, or typographic errors, and many instances of homoiteleuton.

McGovern promised to prepare a critical edition of the text, based on all manuscripts, but none was ever published (thesis, p. xii).

25 Hanna and Griffiths, p. 227.

26 See above, pp. 9-10. Ms L would also not be useful as a base text since it now begins in the middle of Chap. 37.

27 Hanna and Griffiths, p. 227.

28 Ff. 33r–110v: Pèlerin de Prusse, *Liber de astrologicus*; ff. 111r-118v: Pèlerin de Prusse, treatise on the astrolabe; ff. 119r-152v: Pèlerin de Prusse's French version of Johannes Hispalensis' Latin translation of al-Qabīsī (Alcabitius), *Liber introductorius minor*; added material including horoscopes of the king and his children.

an author/presenter (Pèlerin de Prusse?). The identification of these images as Charles is supported by many other scholars.[29] It should also be noted that this manuscript remained in royal hands, and was described in the inventory of books in the royal study at Vincennes in 1418.[30]

As a presentation copy it should best conform to the original version by Oresme. Obviously there was a precursor text, since there are a few minor errors in ms E, but that earlier original text seems not to have survived leaving E as the preferred manuscript.

As for other manuscripts, ms C is a close copy to E, and is also written in a textura rotunda hand but it lacks the flourishes of E and so is less likely to be a presentation copy. There are a few insignificant differences in vocabulary and spelling, and it is probable that it was either a copy of ms E or of the earlier original text from which E might also have been copied. As well it seems to have followed a different path and ended up in the library of Jean Lebégue (1368-1457) and not within any royal collection.[31]

Mss H and N, which form a small sub-set of the other manuscripts, show variations which make either of them unlikely to be the original text – material added, material omitted, material changed.[32] Both manuscripts are written in a relatively clear book hand.

Furthermore, as is obvious from apparatus criticus, the text found in ms E, and as edited here, is much "cleaner" than the other texts, without extraneous explanations, and without serious error.[33]

For all these reasons I am confident that the choice of ms E as the base text for this edition is the correct choice.[34]

---

29 Hanna and Griffiths (p. 229) list a number of studies which link the images to Charles V. This identity has also been supported by Laird, p. 168.

30 Ibid., p. 230.

31 Ouy, pp. 161-162.

32 A sampling: line 110 (*add.* ne son); line 259 (*add.* ainsi sont .xii. parties); lines 620-627 (*om.* Item ... resconsent); line 979 (*add.* car deux quars valent demie heure); line 1193 (au dehors] au dessoubz).

33 As further evidence for the selection of E (and the close copies C and P) is the failure of the other manuscripts to properly understand the phrase "vers et creps fors" (line 1306). See the note to that line.Orléans, Bibliothèque Municipale

34 McGovern (p. xii) also concludes that ms E would be the closest to any original copy.

4
EDITORIAL APPARATUS

Editorial Principles

As stated previously (pp. 16-17), the base text of this edition is ms E (Oxford, St. John's College, ms 164, ff. 1r-32r). It appears to be a presentation copy and perhaps the original. In any event, ms E at least represents a very early copy, with its structure, vocabulary, and spelling. (The folio numbers in the margin are those of ms E.)

Thus the text printed here has tried to follow the base manuscript closely. Where there has been an obvious error, the correction is entered in square brackets with the original noted in the apparatus. Trivial errors have been corrected but the original is still noted.

In ms E, the script sometimes does not distinguish between "n" and "u" as in "on" and "ou" or within other words. I have depended on the other mss or on the sense of the sentence to make my choice.

There are a few common abbreviations in the text, mostly short-cuts found as well in both Old French and in medieval Latin, with which all scribes would be familiar. These include "ō" for "om/on" and "ē" for "em/en". The superscript apostrophe signifying "er" (and sometimes "ir") appears in the text as well as the underscored "p" for "per" or "par". These abbreviations have been expanded without comment.

Apostrophes have been inserted in the text where needed, e.g., "quil" becomes "qu'il" and "cest" becomes "c'est" (except where "cest" is a variant of "ce/ces"). (However, "que il" in ms E is left that way in the text – e.g., line 1356 – and the substitution of "qu'il" for "que il" in some mss is not noted.)

Where words beginning with a vowel are preceded by "d" (for "de") or "l" (for "le" or "la"), an apostrophe has been added to show elision as in modern French. Thus "lespere" is transcribed as "l'espere". Where the "de" or the "le/la" appears before a vowel, that form has been maintained with a word-space between it and the following word.

Consonant / vowel "i" has been distinguished with the use of "j" and "i"; the same for "u" distinguished as "v" or "u".

I have not hyphenated any of the words in the text (hence the uneven spacing within the lines) so that there will be no confusion in the spelling and/or combining of various pairs of words.

Accents follow the rules laid out in *Romania*, 52 (1926)[1] for Old French, i.e., the use of the acute accent on "e"s to denote a final tonic, as well as on the "e" of the second person plural ending, since scribes use "-s" and "-z" interchangeably. The diaeresis and the c-cedilla have not been used.

There is minimal punctuation in the manuscript, other than end of sentence periods or periods where we would use commas, so I have added more modern punctuation in some places in order to aid the reader in understanding the text. This includes more commas, some semicolons, and some dashes. As well some pairs of "sentences" have been joined where the meaning demands it.

References to the various sources mentioned by Oresme will be found in the special "Source References" appendix at the end of the text.

Numerals pose a special problem. Ms E uses roman numerals as do most other manuscripts, although a few substitute arabic numerals (but not always). The numbers in roman numerals are set off with periods, e.g., ".xlv." for "45". In complicated higher numbers, where superscript "c" for hundred(s), "m" for thousand(s), "xx" for 20s are sometimes used, I have tried to present the number as found in the text, e.g., ".ii^{m} v^{c} v." for "2505" (line 769). The variations of the text version in other manuscripts are noted (as are the substitution of words for numerals – "deux" for ".ii." or "2") for those interested in the change-over of roman numerals to arabic numerals. (I have also retained the common medieval use of "iiii" for "iv".)

The Apparatus

Only major variants are included in the apparatus. Spelling variants of the same word are usually not noted (unless the change might lead to a different sense), nor are the changes from copy to copy of double consonants and single consonants. The substitution of "z" for "s" and vice versa (or "x" and "s") is not noted except when a word appears in the apparatus for another reason.

Certain words or letters are unreadable in some manuscript copies (especially D and M). These have been marked in the apparatus criticus as illegible ("*illeg.*"). A dubious reading is marked as "(?)", e.g., line 184 apparatus; an uncertain ending is marked with a swung-dash, as in line 183 apparatus ("mouvon~").

In producing the text it has been necessary to balance the needs of Middle French philologists who wish to see the original spelling of words in the dialect of the text, and the historians of science who wish to have a clear meaning. In the event where a dialect word might lead to confusion in meaning, the word in question has been "translated" into more common French in a second apparatus below the text. In some cases the grammar has not been corrected since the meaning is obvious; thus "il" (line 104) is left in the singular even though the verb ("sont") is plural.

---

1 "Compte Rendu de la Séance tenue a Paris les 18 et 19 octobre 1925": "Establissement de règles pratiques pour l'édition des anciens textes français et provençaux", *Romania* 52 (1926), 243-249.

Minor changes in word order are, for the most part, not noted if all the words appear in both the main text and the copy. (However, if the change in word order also changes the meaning, the change is noted.)

Diagrams

The diagrams in the text are those found in the manuscripts, based on the ones in ms E. Where additional information has been added to the diagrams in other manuscripts, this is presented in italic type, with indications in the apparatus criticus for the diagrams of where this information appears. (Note: only the manuscripts in which the additional information appears are listed, and not those from which the diagram is omitted.)

The apparatus criticus (where there are no line numbers) generally runs top-to-bottom and left-to-right.

The diagrams are numbered according to the chapters in which they are found – thus there is no continuous numbering. Also included is information indicating which manuscripts a particular diagram is found in. (The same system applies to the charts.)

If there is a significant different version of the diagram in some manuscripts, a second, alternate diagram has been included.

Nicole Oresme

Traitié de l'espere

Sigla

A Bern, Burgerbibliothek, ms 310, ff. 1ra-27r
Begins Chap. 12, line 354.

B Bordeaux, Bibliothèque municipale, ms. 531, ff. 90r-127r

C Firenze, Biblioteca Medicea Laurenziana, ms. Ashburnham 1604, ff. 1r-39v

D Leiden, Bibliotheek der Rijksuniversiteit, ms. Vossius gall. folio 10, ff. 1r-31v

E Oxford, St. John's College, ms. 164, ff. 1r-32r
My base manuscript.

F Paris, Bibliothèque nationale de France, ms. fr. 565, ff. 1r-22v

G Paris, Bibliothèque nationale de France, ms. fr. 1083, ff. 126ra-145rv

H Paris, Bibliothèque nationale de France, ms. fr. 1350, ff. 1ra-38v

K Paris, Bibliothèque nationale de France, ms. fr. 2240, ff. 61r-96r

L Paris, Bibliothèque nationale de France, ms. fr. 24278, ff. 140r-146r
Begins Chap. 37, lines 1150-1051.

M Paris, Bibliothèque nationale de France, ms. nouv. acq. fr. 1052, ff. 1r-38r

N Paris, Bibliothèque nationale de France, ms. nouv. acq. fr. 10045, ff. 1r-39v

P Vatican, BAV, Reg. lat. 1337, ff. 29r-44v

f. 1r / Ci commence *L'espere* en francois.[1]

LE PROLOGUE

La figure et la disposicion du monde, le nombre et ordre des elemens, et les mouvemens des corps du ciel appartienent assavoir a tout homme qui est de franche condicion et de noble engin, et est chose bonne et bele assavoir et delectable et profitable et honneste. Et avequez ces, est neccessaire pour savoir philosophie, et par especial pour astrologie.

Mes afin que engin humain peust plus legierement tele chose comprendre, les sages anciens composerent entre les autres un instrument qui est appelé "espere materiel" ou "artificiel", lequel on puet resgarder tout entour manier et tourner et y considerer en partie la descripcion et le mouvement du monde et du ciel aussi comme en un examplaire – duquel je vueil dire en francois generalment et plainement ce qui est couvenable pour savoir a tout homme sans moi profunder es

1 *before* Ci] Domine patris et filii et spiritus sanci, Amen C L'espere] De spera HN francois] *add.* que translata Maistre Nichole Oresme N; *add.* que translata Maistre Nicole Oresme, tres excellent philosophe, et c. H 1-2 Ci ... prologue] *om.* BDFG; In nomine Domini, Amen. Cy commence l'espere translatee de latin en francoys par Maistre Nicole Oresme, et premierement le prologue s'ensuyt M; Ihehis - Maria | Cy commence le tractiés de l'espere que translata de latin en francois tres excellent philosophe Maastre Nychole Oresmes K 2 Le prologue] *om.* HN 3 et_1] de P elemens et les] *om.* B 4 du ciel] celestes M assavoir] assavoir *corr. to* a savoir C 5 chose ... et_4] belle chose BDFGHKMN assavoir] a savoir C 6 profitable] pourffitable D ces] *om.* N philosophie] ph'ie EFP 7 pour] *om.* BM; *add.* savoir N 8 afin] *om.* K; a celle fin D 9 entre les autres] *om.* M un] .i. G qui est] *om.* KM appele] *add.* le H 10 lequel] ouquel M on] en G manier] mouvoir F y] *om.* N 11 aussi] ainsi M 12 un] 1/.i. CEGP; une HN 13 plainem ent] premerement H qui] *add.* en H couvenable] profitable B pour] a H; de M pour savoir] assavoir BKN sans moi] en *illeg.* D profunder] *corr. from* parfonder N

---

1 In several mss the text begins with a large image: In ms E, there is a crowned figure (King Charles V) seated with a book on his lap, with a book case behind him and a table in front on which there are more books and an armillary sphere; at the beginning of Chap. 1 in the ornate initial letter "L", there is an image of Oresme standing and holding an armillary sphere (Hanna and Griffiths, p. 229, suggest that he is dressed as a Franciscan !). In ms C, at the beginning of Chap. 1, there is an image of "Oresme" seated and also holding an armillary sphere; in ms H, at the beginning of the prologue, there is a similar image of Oresme but also with several pupils(?) standing and listening. In ms F, at the beginning of the prologue, Oresme is seated, writing, at a desk with an armillary sphere (on a stand) in front of him.

In his 1974 thesis, McGovern did not recognize that lines 1-2, found in red in a three-line area at the right end of the first lines of the folio, was the title, and integrated the various words into the text of lines 3 and 4.

demonstrations et es subtiletés qui appartiennent as astrologiens. Et vueil diviser cest oeuvre par chapitres.

CI SONT LES CHAPISTRES, LE PREMIER[2]

14 *and elsewhere* es] *read* les
30, 31 ozison / ozizon] *read* (h)orizon[4]

14 et_1] ne N es] *om.* B as] aux BDGKMP; es N 14-15 cest oeuvre] *om.* B; ce livre M; euvre N 16 Ci ... chapitres] *om.* BF; Ci s'ensuit l'ordre des chapitres de ce present livre K; Cy apres s'ensuyient les chapitres par ordre M; Cy commencent les chapitres HN; Cy sont les chapitres par ordre G 16-68 Ci ... oeuvre] *om.* D 17 .i.] Le premier chapistre B ses] ces F principaus] *om.* M 18 .ii.] Le secont chapistre est B 19 .iii.] Le tiers chapistre est B 20 .iiii.] Le quart est B 21 .v.] Le quint B et de l'equinocial] *om.* BK 22 .vi.] Le .vime B 23 .vii.] Le .viime B 24 .viii.] Le .viiime B 25 .ix. Le .ixme B 26 .x.] Le .xme B la] *om.* G 27 .xi.] Le .xime B signe] figure H prins] *om.* F .iii.] troiz/trois/troys FMN; 4 P 28 .xii.] Le .xiime chapistre est B Des] de H .ii.] deux BFHMP 29 .xiii.] Le .xiiime B 30 .xiiii.] Le .xiiiime B De] Du K 31 .xv.] Le .xvme B De] Des M .ii.] deux BFHKMNP

---

2 For a description of the layout of the chapter titles in each ms, see the Addenda to the Prologue.

3 I have added the beginning line number for each Capitulum, to act as a table of contents although such a cross-reference does not exist in any of the manuscripts.

4 "Ozizon" is an odd variant for "horizon" (late Latin and French) but is used by ms E throughout the text. The only other attestation to this spelling is also in Oresme (*La Livre du ciel*, Bk. II, c. 6, line 11 [p. 324]). This may simply be an error by one scribe, since the medieval letter "r" is often written in a fashion similar to the letter "z" – "z" is written in the form of a "3" and "r" as "2" (or "3" with a shorter tail). In this treatise, "r" as "2" appears in many words: "ap2ouver" (l. 79), "fo2s" (l. 90), "to2ne" (line 105), "co2ruptibile" (l. 123), and so on. These forms of "z" replacing "r" in "ozizon" also appear in mss C and P and some of the other mss (H and K), but not in BFGMN.

32 .xvi.] Le .xvi.^{e} B l'elevation] la levation BG 33 .xvii.] Le .xvii.^{e} B Des] De F .iiii.] quatre MP; troys B; troiz F; .iii. G 34 .xviii.] Le .xviii.^{e} B 35 .xix.] Le .xix.^{e} B 35-36 droit ... ozizon] *om.* P 36 .xx.] .xxi. E; Le .xxme B De] Du M 37 en] *om.* F; en de EP; est B en l'espere] espere KM 38 .xxi.] .xxii. E; Le .xxime B Des] De FGK .ii.] deux BFHMP levement] *add.* et resconsement M 39 .xxii.] .xxiii. E; Le .xxiime B 40 .xxiii.] .xxiv. E; Le .xxiiime B grant] partie ou B 41 .xxiiii.] .xxv. E; Le .xxiiiime B l'inequalite] la nequalite B 42 .xxv.] *om.* E; Le .xxvme B .xxv. D'une ... naturels] *om.* F; D'une aultre cause de l'inequalité des jours M(*this line was skipped and then added after line 68*) cause de] *om.* FHKN inequalite] equalite K; cause de la nequalite B; cause de l'inequalite CG 43 .xxvi.] Le .xxvime B 44 .xxvii.] Le .xxviime B qui] *add.* y H 45 .xxviii.] Le .xxviiime B en] *add.* .xxix. E .v.] cing/cincg BM 46 .xxix.] .xxx. E; Le .xxixme B 47 .xxx.] .xxxi. E; Le .xxx.^{e} B .xxx. ... especial] *om.* FN 48 .xxxi.] *om.* E; Le .xxxime B De$_1$ la] La CP de la terre] *om.* B selonc] secun C; *add.* les H aucuns] anciens HN; austuns M; enciens K 49 .xxxii.] Le .xxxiime B La] De la BFGHKMN la$_2$] *om.* K astrologiens] astrologes F 50 .xxxiii.] Le .xxxiiime B 51 .xxxiiii.] Le .xxxiiiime B la] *om.* G 52 .xxxv.] Le .xxxve B chascun] ceste F 53 .xxxvi.] Le. xxxvie B .xxxvi. ... climat] *om.* F l'elevacion] le levacion *some*; la levation BG en] en en E

5 The scribe of ms E most often writes "cecle" instead of "cercle" throughout.

.xxxvii. De la quantité de terre habitable. ⟨1140⟩
.xxxviii. Des habitations qui sont dehors les climats. ⟨1168⟩
.xxxix. D'une merveilleuse consideracion ou circuite de la terre. ⟨1210⟩
.xl. Des causes pour quoi aucun leu est habitable ou non. ⟨1259⟩
.xli. Des extremités de terre habitable. ⟨1294⟩
.xlii. De la difference des particz bien habitables selonc leur latitude. ⟨1316⟩
.xliii. De la difference des partiez bien habitables selonc leur longitude. ⟨1344⟩
.xliiii. Du croissement et appetissement de la lune. ⟨1369⟩
.xlv. De [la] declaracion plus plaine des choses dessus dictes. ⟨1408⟩
f. 2^{r} / .xlvi. Des causes des eclipses en general. ⟨1452⟩
.xlvii. De escl[i]pse du solail en especial. ⟨1479⟩
.xlviii. De l'eclipse de lune en especial. ⟨1504⟩
.xlix. Des couleurs de la lune en temps d'eclipse, et des eclipses des estoiles. ⟨1531⟩
.l. La fin de ceste oeuvre. ⟨1560⟩

54 .xxxvii.] Le .xxxvii^{e} B de$_2$] de la KM 55 .xxxviii.] Le .xxxviii.^{e} B dehors les] hors des M les] des P 56 .xxxix.] Le .xxxix.^{e} B
57 .xl.] Le .xl.^{e} B aucun] *om.* B; un/ung FHKN; .i. G aucun ... habitable] aulcuns lieux sont habitables M 58 .xli.] Le .xli. B de] de la KM 59 .xlii.] Le .xliime B .xlii. ... longitude] *chapter headings .xlii. and .xliii. inverted, inversion corrected by use of marginal "b, a"* M selonc] secun C 60 .xliii.] Le .xliiime B selonc] secun C 61 .xliiii.] Le .xliiiime B appetissement] descroissement K 62 .xlv.] Le .xlvme B De la] *om.* F la] *om.* CEGP declaracion] eslevacion B plaine] ample M(*illeg.; from McCarthy*) dictes] dcēs E 63 .xlvi.] Le .xlvime B 64 .xlvii.] Le .xlviime B De esclipse] De esclpse E; D'eclipse F; De l'eclipse HM; Des esclipses K du] de FKN 65 .xlviii.] Le .xlviiime B De l'eclipse] D'eclipse FG; De eclipse N; Des esclipses K de$_2$] de la FGKM 66 .xlix.] Le .xlix.^{e} B en] ou N d'eclipse] de eclipse P; de l'eclipse N des$_2$] *om.* P et des eclipses] *om.* B; et de eclipse H; et de l'esclipses G; et d'esclipse K; et de l'eclipse N 66-67 et ... estoiles] *om.* M
67 des estoiles] d'estoilles K 68 .l.] *om.* EP; Le .l.^{e} est B .l. ... oeuvre] De l'eclipse des estoilles 50 M; *add.* etc. K oeuvre] euvre N; *add.* L[*illeg.*] doit estre ung [*illeg.*] qui tient une espere de ceste figure P[6]

Addenda: Prologue

Ms E (which in this section of the text has the chapter titles in two columns) has the chapter numbers (in red roman numerals) at the end of the first line of each chapter title, even if the title extends to a second or third line. Chapters 20 and 28 are doubled numbered (on their second line) by the rubricator which throws off the

6 This line in ms P precedes Chap. 1, and the small figure of an armillary sphere.

numbering of the following chapters, and which is only corrected when the numbering is skipped for Chapters 25 and 31.

In ms B, the chapter titles begin with "The first ...," "The second ...," etc. as recorded in the apparatus criticus.

In ms C (the chapter titles in lines 17-31 are in two columns) the titles are numbered in roman numerals down the right-hand side of each column/page.

The list of chapter titles is missing in ms D.

In ms F (the whole text is in two columns), the chapter titles are not numbered.

In ms G (the whole text is in two columns) the chapter titles are numbered down the left-hand side using the late medieval versions of the arabic numerals.

Ms H (the whole text is in two columns) has the numbers for the chapters (in roman numerals) in the left or right page margins beside the titles.

Ms K (the text is in one column) has the numbers for the chapters (in roman numerals) down the right-hand side of the page.

In ms M the chapter titles are numbered to the right using arabic numerals.

In ms N the chapter titles run on from line to line with no numeration.

Ms P (again with the chapter titles in two columns) has the numbers for chapters 1 to 30 (using arabic numerals) at the end of the titles. The rest of the titles are not numbered.

LE PREMIER CHAPISTRE DE LA FIGURE DU MONDE ET DE SES PARTIES PRINCIPAULS[1]

Le monde est tout roont aussi comme une pelote. Et les philosophes tout corps de cele figure appellent "espere" ou "chose sperique". Et dient que l'espere de monde est composee du ciel et des quatre elemens comme des .v. principauls parties.

Premierement est la terre toute masseice et roonde non pas parfectement, car il y a montaignez et valeez, mais toutesvoies tent elle a roondesce. Et qui seroit ou ciel et la resgarderoit, elle sembleroit estre toute roonde.

Et pour ce, l'eclipse de la lune apert en roondesce, laquele eclipse est causee de l'ombre de la terre, si comme il apparoistra aprés.[2] Ne je ne me vueil plus arrester a prouver que la terre est roonde. Et[3] est la terre ou centre et ou milieu du monde, car ce est le plus pesant element. Et est assavoir que le milieu d'une esp[e]re ou d'un cercle est apelé "centre"; et chascune droite ligne qui passe au travers de la terre, ou de quelconques espere ou cercle, parmi le centre de cele espere ou de ce cercle, est appelée "dyametre" de celle chose; et la motié du dyametre qui part du centre est f. 2v appellé "semidyametre". /

Aprés la terre est l'yaue ou la mer, mes elle ne cuevre pas toute la terre. Car aucune partie de la terre n'est pas de si pesante nature comme l'autre, aussi comme

69 Le ... chapistre] *om.* HKN; *add.* est BDG 69-70 Le ... principauls] *om.* F; Cy apres s'ensuyt le premier chapitre qui est de la figure du monde M; *add.* .i. cap^{m} H; *add.* .i. K; *add.* .i.^{e} c N 71 aussi] *om.* B philosophes] *add.* dient que N tout$_2$] tout tel K corps] *add.* ront] H de] du M 71-72 de cele] de tele/telle CGHN; et telle K 72 appellent] est appelle N 73 quatre] .iiii. FGKN; 4 D des$_2$] de BCP .v.] cinq/cincq BFMP; 5 D 74 Premierement] Primo K 75 mais] *om.* M tent] *om.* M Et$_2$] que B 76 resgarderoit] en regardant M estre] *om.* BDFGHKMN 77 eclipse$_2$] chouse G 78 apparoistra] apparra BFGHKMN je] plus M plus] *om.* M; pas BG arrester] meptie(?) B 79 Et$_1$ ... terre$_2$] La terre si est M ou$_2$] et au P; et ou CFGKMN 79-80 car ce] qui D 80 espere] espoire E 80-81 ou ... cercle] *om.* B 81 qui passe] passant M 82 quelconques] autre B ou cercle] *om.* B parmi] par B cele espere] d'espere M de ce] du M 83 dyametre] le centre HN 85 mes elle] muelle F 86 terre] *om.* F de$_2$] *om.* B nature] *om.* B; *add. illeg.* D aussi] ainsi M

---

1 For the images found at the beginning of Chap. 1 (mss EC), see the footnote at the beginning of the Prologue (line 1). Also, in ms P, there is a small image of an armillary sphere set here at the beginning of Chap. 1.

2 Chap. 48, ll. 1505 ff.

3 Oresme (lines 79-107) presents the traditional Aristotelian four-element sensible world, centred on the earth, covered with water, then air, then fire, beyond which is the region of the planets and stars.

nous voions que estain ne poise pas tant comme plonc. Et pour ce la partie mains pesante est plus haute et plus loing du centre et descouverte d'yaue, afin que les bestes y puissent vivre. Et est aussi comme la face et le visage de la terre tout descouvert, fors que parmi ha aucunes petites mers, ou bras de mer, et fleuves, et tout le demourant est aussi comme enchaperonné, vestu et affublé de la grant mer.

Aprés est l'espere de l'air, qui enclot terre et mer, et est devisee en trois regions, dont l'une est basse la ou habitent bestes et oisiaux. L'autre, moienne ou sont les nues et teles impressions, et est touzjours froide. La tierce region est plus haute la ou il ne ha vent ne pluie ne foudre ne autre tele impression. Et sont aucunes montaignes qui attiennent jusques la, si comme est Olimpus.

Aprés, selonc les philosophes, est l'espere du feu, et n'est pas flambe ne charbon, mais est feu pur et [in]visible pour sa tres grant clarté. Car tout aussi comme
f. 3r l'yaue est plus clere que la terre et comme / l'air est plus cler et plus legier que l'iaue, aussi est cil feu plus cler et plus legier que l'air. Et autre tant, ou plus, est le ciel cler que n'est le feu en son espere excepté les estoiles qui sont les plus espesses partiez du ciel. Et pour ce sont il visible et luisans. Et dit Aristotes[4] que en la plus haute region de l'air, pres de l'espere du feu sont les estoiles cometes, quant il sont, et que l'espere du feu et celle haute region de l'air et les estoiles cometes tout ensemble torne aveques le ciel d'orient en occident, aussi comme se le ciel traysist tout ceci avec soi par son mouvement. Et par dessus touz les .iiii. elemens est le ciel.

87 plonc] *read* plomb *or* plonb

87 estain] lestaing B estain ... plonc] estaing n'est pas si pesant que plomb M plonc] le plomb B; plom D; plon GHN; plont C; plum F ce] *om.* F 88 pesante] pondereuse M et$_2$] et est 89 Et est] *om.* D aussi] ainsi M et$_2$] il come' est il scondte aussi est (?) D 90 ou] *om* BHK 91 aussi ... affuble] couvert d'eaue ainsi comme enchaperonnee vestue et affublee M comme] *om.* G et affuble] *om.* B 92 trois] .iii. FGHK; 3 D 93 moienne] *add.* la FGM sont] habitent M 93-94 les nues] mines F; ers nues K 94 teles] estoilles B 95 ne ha] n'a ne GHK; ne a ne F; n'y a ne M autre ... impression] autres telles impressions M 96 attiennent] *add.* de cy HK jusques] *om.* D; siques CFG si] aussi N est] *om.* DN; *add.* le Mont M 97 selonc] secon/secou C ne] ou 98 charbon mais est] *om.* K invisible] visible EP Car] que D aussi] ainsi FM 99 comme$_2$] *om.* B est$_2$] *om.* B et plus legier] *om.* B 99-100 est$_2$ ... l'air] *om.* K 100 aussi] ainsi M cil] icellui P; il F; le BMN plus$_1$] *om.* D plus$_1$ ... legier] *om.* B que] *add.* ne est/n'est DFG autre tant ou plus] tout ainsi B 101 ciel] *add.* de ciel en ciel plus HK; *add.* plus BCDFM le$_2$] *om.* B en] ou DH 102 Et$_2$] Ce C Aristotes] Aristote H 103 de$_2$] *om.* B 104 quant il sont] *om.* N il] ilz BDHKP; *illeg.* M les estoiles] cellez B 105 torne] tournent FHK aussi] ainsi M se] *om.* BG 105-106 aussi ... soi] et K 106 traysist] *add.* a soy H traysist ... ceci] tyrast tout ce avecques soy M avec soi] *om.* H; a soy N son] sont B .iiii.] quatre BHM; 4 D

4 See "Source References," line 102.

DIAGRAM: CHAPTER 1

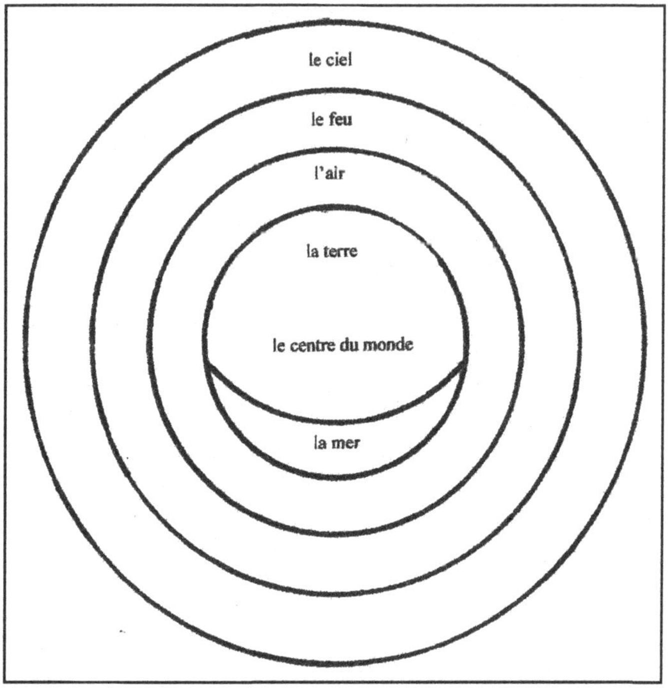

Complete diagram: E CP
No diagram (or space only): BDFGHKMNP

LE .II.E CHAPISTRE. DE LA NATURE DU CIEL

Le ciel ne est proprement ne pesant, ne legier, ne dur, ne moul, ne cler, n'espés, ne chaut, ne froit, ne si n'a saveur, ne oudeur, ne couleur, ne teles qualités fors que il est chaut en vertu. Car il puet causer chaleur ici bas par sa lumiere, par son mouvement, et par son influence, et est improprement dur. Car [il] ne puet estre divisé ne cassé. Et aussi est il improprement colouré de lumiere en aucune partie et espés ou dempse. Car l'estoile est la plus espesse partie de son espere ou de son ciel.

Et si n'y puet estre estoilé ou autre partie adjoustee ou ostee, ne ne puet croistre, ne apeticier, ne estre d'autre figure, ne muer, ne changier, ne empirier, ne enveillir, ne estre corrompu, ne alteré en quelconques maniere fors aucunement en lumiere tant seulement si comme il apert en l'eclipse de la lune. Et si ne puet estre arresté ne reposer ne torner d'autre cause, ne plus tost, ne plus tart, ne en tout, ne en partie, ne soi avoir autrement que selonc son commun cours ordené, se ce ne estoit par miracle divin. Et pour ce est le ciel et les estoiles d'autre nature et d'autre maniere que les .iiii. elemens et les choses qui en sont composee. Car tout ce est transmuable et corruptible; et Aristote[1] dit que toute chose corruptible sera aucune fois corrompue selonc le cours naturel.

108 Le ... chapistre] *om.* HKN; Le second chapitre M; Le secont chapistre est BG ciel] *add.* .ii. K; *add.* .ii. chapitre HN 109 ne$_2$] *om.* M ne$_{4, 5, 6}$] *om.* B 110] ne$_2$] et M n'a] en H; ne F; *add.* ne G couleur] *add.* ne son BCDFGHKMN 111 Car] que D il] *om.* E par son] *om.* B 112 il] *om.* EP 112-113 dur ... improprement] *om.* D 114 ou dempse] condempsé K Car] que D 115 n'y] ne BDMN estre] *om.* DH partie] *add.* estre DH ou ostee] *om.* D; ou soustraiete M 115-116 croistre ... ne$_4$] *om.* M 116 ne$_{1, 2}$] ou H enveillir] vieillir P 118 Et si] Et s'il M 119 torner] retorner D cause] guise DFGHN 120 soi] se B selonc] secon/secou C commun] premier N cours ordene] cours N; mouvement B se ce ne estoit] *illeg.* D 121-122 et$_3$ .. que] que me soy B 122 que] *add.* ne sont M .iiii.] 4 D; quatre BHM et] *add.* dont D Car] *illeg.* D transmuable] muable D 123 Aristote] Aristotile B aucune] une M 124 selonc] secon/secou C naturel] de nature BCDFGHKN

1 See "Source References," line. 123.

LE .IIIE CHAPISTRE. DES PARTIES DU CIEL

Toute la masse des .iiii. elemens et des choses qui y sont est appelee "l'espere des choses corruptibles", et est enclose ou ciel aussi comme le moiel de l'uef est enclos en l'aubun. Et tout le ciel qui est par dessus est devisé en plusieurs parties ou esperes. Premierement au plus pres des elemens est l'espere de la lune, qui enclot en soi sans moien l'espere du feu. Et le corps de la lune que nous voion est contenu et fichié
f. 3^{v} de/dens cele espere, et est meu du mouvement d'elle aussi comme .i. clou qui est fichié en une noif. Et ainsi sont les autres planetes et estoiles fichiés en leur esperes.

Aprés dessus est l'espere de Mercure; aprés, l'espere de Venus; puis, l'espere du soleil; puis, celle de Mars; aprés, celle de Jupiter; et aprés est l'espere de Saturne. Et ainsi sont les esperes des .vii. planetes selonc leur ordre.

En oultre par dessus est le ciel estelé, le[1] octave espere, ou sont les "estoiles fichiés" qui sont ainsi appellees non pas pour ce que les autres ne soient fichiés chascune en son espere, mes pour ce qui sont meues plus regulierement, et toutez d'une guise, et ensuivent plus que les autres.

Le premier mouvement: Puis selonc les astrologiens, est par dessus tout, la nuefvieme espere ou il n'apiert nule estoile. Et la cause pour quoy il dient que elle y est, si est pour ce que il appiert en le .viii.^{e} espere plus d'un mouvement; si couvient qu'il en ait une par dessus qui soit meue tant seulement d'un simple mouvement.

125 Le ... chapistre] *om.* HN; *add.* est B .iii.^{e}] tiers BFGM; 3^{e} P ciel] *add.* .iii. K; *add.* .iii. ca. H; *add.* .iii. chapitre N 126 .iiii.] quatre BGMN; 4 D 127 aussi] ainsi M moiel] mouel HN; moyen M 128 l'aubun] *add.* d'icellui M pardessus] dessus K ou] en G 129 Premierement] Primo K plus] *om.* B des] de K; *add.* 4 P 131 cele] son HN meu] meue M .i.] un/ung BDFHKMNP 132 en$_{1}$] dedans M ainsi] aussi N et$_{2}$] *add.* les G leur] leurs B 133 dessus] audessus D puis] apres G 134 celle$_{1}$] l'espere D apres$_{1}$] puis HN celle$_{2}$] *om.* B apres$_{2}$] puis M l'espere] celle BD 135 .vii.] *om.* BDFGHKN; sept MP selonc] secon C 136 En oultre] Et oultre K; Et oultre et M estele] estoille N le octave] le huitiesme P; le .viii.^{e} BHKN; qui est la huytieme M; *corr. to* huicttesme D 137 pour ce] *om.* P ne soient] *om.* M; *add.* pour tamt P 138 qui] que elles HN; *add.* il/ilz DG; *add.* les P et toutez] *om.* F 139 guise] cause P 140 Puis] *ms* E *begins the new paragraph here* selonc] secon/secou C 141 nuefvieme] novisme DFG; .ix.^{e} BHK il$_{2}$] ilz HK elle] icelle ix^{e} espere K 142 y] *om.* M il ... espere] en huitiesme espere appiert K appiert] voient M .viii.^{e}] huitiesme BMP; 8^{e} D 142-143 si ... par dessus] Parquoy fult qu'il y ait ung aultre movement pardessus ou une aultre espere M 143 ait] yait GKN tant seulement] *om.* HKN

---

1 While "espere" is treated as feminine, there are times (such as here) where it is preceded by the masculine article. This has been retained.

Encor dient aucuns que par dessus a .i. ciel immouvable; puis, .i. ciel de cristal; et puis le ciel empyree ou le throne de Salmon. Et teles choses qui ne appartienent pas a naturel philosophie ne astrologie. Pour quoy il souffist a present parler de .ix. esperes dessus ditez. Et fait bien a noter la merveilleuse distribution de ces estoiles. Car en la .ix.^{e} et souveraine espere ne a nule; et en le .viii.^{e} en a tant que on ne les puet nombrer; et es autres en descendant en a en chascune une, tant seulement en appelant estoiles, le soleil et la lune.

145 .i.$_1$] un/ung DFHKMNP .i.$_2$] le F; un/ung DHKMNP ciel$_2$] *add.* immouvable M 146 empyree] empireal M ou] *add.* est N Salmon] Salomon FGHMN; Solomon K Et teles] ou semblable M 147 a ... philosophie] aux naturels philosophes M ne] *add.* a HKN astrologie] *add.* sont K il] *om.* K il ... present] pour le present se fault arrester et M de .ix.] de .iv. F; de neur P; des .ix.^{e} K; des neuf M; de 9 D 148 fait] est M de] des F ces] ses N 149 .ix.^{e}] neufmesme MP; *corr. to* neufiesme D et$_1$] qui est D et$_1$ ... et$_2$] n'a nulles estoiles mais M espere] *om.* BFN ne$_1$] n'en BHKN le] *om.* D on] nul home FG; nul ha BD ne$_2$] *om.* P 150 en a] et H une] espere vue estoille N

LE QUART CHAPISTRE. DE LA FIGURE DES ESPERES DU CIEL

Chascune de ces esperes est gradement espesse ou parfonde, car l'espere de la lune atteint des l'espere du feu siques a l'espere de Mercure; et l'espere de Mercure jusques a celle[1] de Venus; et ainsi des autres en .i. montant, car il n'y puet avoir riens veu. Et est chascune espere d'une meisme espesseur ou parfondeur en ses parties tout environ en chascun endroit.

Et a chascune une face ou superfice dehors ou dessus qui est appellee "superfice convexce". Et ha aussi une autre face ou superfice dedens ou dessous qui est appelee "superfice concave".

Et l'une et l'autre de ces superfices est sperique ou roonde et polie en tele maniere que on ne pourroit ymaginer chose[2] ou monde plus roonde ou miex tornee f. 4^{r} ne miex polie. / Et est chascune concentrique au monde; c'est a dire, que le centre et le milieu du monde, ou le centre de la terre, est milieu et centre de chascune de ces esperes du ciel, et de leur superfices dessus dites.

Et chascune de ces esperes touche et enclot ou environe une autre, et est enclosé d'une autre excepté la premiere et la derreniere, si comme est l'espere de Mercure, qui enclot l'espere de la lune en sa superfice concave, et est enclose en la superfice

152 Le ... chapistre] *om.* HKN; Le quart chapitre qui est M quart] .iiii. DF ciel] *add.* .iiii. K; *add.* .iiii. ca. HN 153 ou] et BM car] que D 154 des] de MKP des ... feu] *om.* D siques] jusques HKMNP de$_{1, 2}$] du F 155 jusques] siques CF celle] l'espere M ainsi] aussi D en] eu E .i.] *om.* BDFGHKMN; sept P car] que D puet] *add.* estre ne BFGHKMN 156 riens] rien de HN veu] de vuit/vieu/viut/vuyt FGKM Et] *add.* combien que l'une soit plus espesse que l'autre, toutes voies BDFGHKN; *add.* combien que l'une soit plus espesse que l'autre en matiere, toutesfois M 158 une] espere une M ou$_1$] *add.* une G 159 ha] *om.* B face ou] *om.* K 159-160 convexce ... superfice] *om.* D dedens ... superfice] *om.* M 161 Et$_1$] *om.* M et$_2$] ou N ces] ses GN 162 on] homme H ne] n'y K ymaginer] *add.* .i. P ou$_1$] au P ou$_2$] ne BDFG tornee] trouvee K 163 miex] *om.* M est] *om.* K et$_2$] est N 163-165 et$_2$ ... dites] d'une cascune espere est ou millieu du monde avec le centre de la gravité de la terre M 164 est] *add.* au H; *add.* ou BDFGKN et] *add.* le BDFGHN ces] ses BDN 165 et de leur] et leurs B; et leurs G; est leurs N 166 de ces esperes] *om.* M; *add.* dessus N; *add.* dessus dites HK ces] ses BGN et$_2$] ou N 167 est] en B 168 enclot] clot K; *add.* de G 168-169 et ... convexe] *om.* N

1 McCarthy reported that in ms G, fol. 127^{v} was blank and that the text skips from here ("celle" – the end of fol. 127^{r}) to fol. 128^{r} (line 197: "partie"). However, there must have been an error in the photographic copy of the manuscript that she was using, since fol. 127^{v} does exist with all the missing text, and is collated here.

2 While P has ".i." before "chose" which makes better sense, C does not. E has "i" at this point but with an erasure dot beneath it, which makes it somewhat unclear what the scribe meant.

concave de l'espere de Venus, laquelle elle touche de sa superfice convexe. Et ainsi des autres, fors la derreniere, qui n'est de nule enclose. Car au dehors par dessus ne ha ne corps ne lieu ne plain ne vuit, selonc ce que dist Aristote.[3]

Et le espere de la lune la ou est le corps de la lune ne enclot nule autre espere du ciel, mes enclot l'espere du feu, et le feu l'air, et l'air l'yaue, et l'yaue la terre, fors tant que l'espere de l'yaue est aussi comme percie la ou la terre est descouverte.

Et l'espere de la terre n'a point de superfice concave aussi comme les autres ont. Car elle est la premiere a commencier embas et est toute masseice, et ha son centre en soy meismes, et les autres sont creuses et ont leur centre en la terre comme dit est.

Telle est la disposicion du monde, et apert ici en figure.

174 percie] *read* partie
177 masseice] *read* massive

169 laquelle elle] laqu'elle B de$_2$] *om.* D sa] *om.* HK ainsi] aussi B 170 fors] *add.* de BDHK 170-171 ne ha] n'a DGHN; ne K 171 ne$_1$] *om.* D selonc] secon/secou C 172 la$_2$ ... lune$_2$] *om.* M la$_2$ ... enclot] n'est enclot D autre] *om.* HKN 174 tant] *om.* D percie] *illeg.* BM; partie FHKN; partiee D 176] aussi] ainsi FM comme] que B 177 ont] *om.* B a] au HKN commencier] commencement K masseice] *illeg.* M; maisture(?) D; massaisce FK; masseisse BHN; massice P 178 creuses] croises K 180 Telle est] Cy ensieuvst B; Et telle G; Et telle est KN et ... ici] *om.* B et ...figure] que s'eust en la figure M ici en figure] ici en ceste figure present le povez veoir H; en ceste figure cy present N

3 See "Source References," line 172.

DIAGRAM: CHAPTER 4

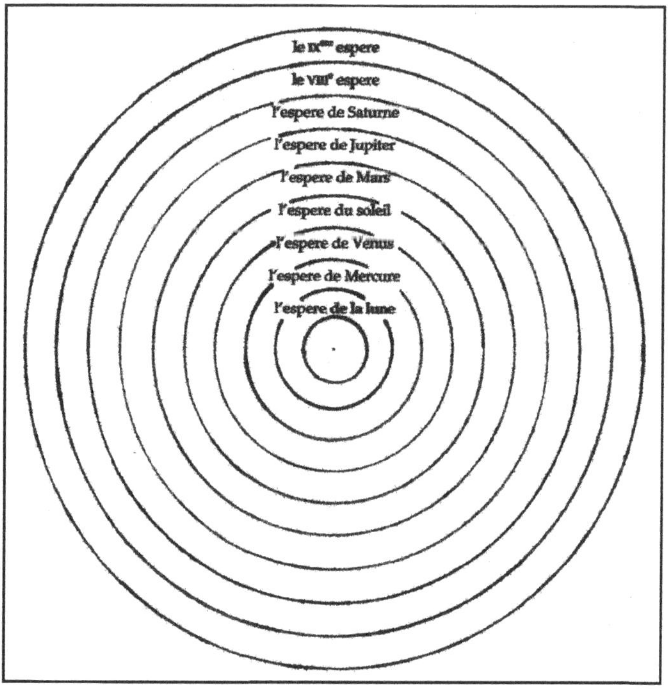

Les .ix. esperes du ciel

Complete diagram: E BCFGHKMNP
No diagram (or space only): D

le .ixme espere] l'espere neufivesme F; la .ix.^{e} espere H; la neufvisme espere B; 10 spera M; .ix. N; la 9me espere P le .viii.^{e} espere] l'espere huitiesme F; la .viii.^{e} H; 9 spera M; .viii. N; le 8me espere P [*inserted circle*] celum stellarum M l'espere de Saturne] Saturnus HMN; Saturnus la 7me espere P l'espere de Jupiter] Jupiter HMN; Jupiter la 6me espere P l'espere de Mars] Mars HMN; Mars la 5me espere P l'espere du soleil] Sol HMN; le soleil la 4me espere P l'espere de Venus] Venus HMN; Venus la 3me espere P l'espere de Mercure] Mercurius HMN; Mercurus la 2me espere P l'espere de la lune] luna HNN; la lune la premiere espere P [*further circles*] l'espere du feu | l'espere de l'aer | l'espere de l'esue | terra B; ignis | aer | aqua | terra | centre du monde H; ignis | aer | aqua | terra M; ignis | aer | terra | aqua N Les .ix. experes du ciel] *om.* HKMP; L'espere des elemens F; Les elemens G; L'espere de la terre | de en fer terre N; Cy enseust la disposition du monde en figure B

f. 4^{v} /

CE EST LE QUINT CHAPISTRE. DE L'ESSEL ET DES POLES DU MONDE ET DE L'EQUINOCIAL

Nous voions sensiblement tout le ciel estre meu de orient e[n] occident, et ce est appellé "le mouvement journel". Et pour ce couvient il ymaginer aussi comme .i. essel environ lequel le ciel tourne, et doit on ymaginer que il passe parmi la terre comme droite ligne et que il ait .ii. bous en .ii. poins du ciel, lesquels .ii. poins ne se mouvent, et sont appelés "les poles du monde", dequoy li un est pres de l'estoile du nort qui touzjours nous apparoist, et est appellé le "pole artique" ou "septentrionel". Car ceste partie du ciel est appellee "septentrion", pour .vii. estoiles notables qui y sont, que les gens lais appelent "le char",[1] et si fait Ptholomee.

Et l'autre pole est a l'opposite tousjours mucié quant a nous. Et est appellé "le pole antartique" ou "le pole austral".

Et l'essel du monde duquel je ai parlé est tant seulement par ymagination et est senefié par la broche qui passe parmi l'espere materiel.

181 Ce est] *om.* B Ce ... chapistre] *om.* HKN; Le .v.^{e} G; Le cinq^{e} chapitre M; .v. chapitre D quint] .v. F 182 des] de K et] *add.* aussi M et de l'equinocial] *om.* BDN; *add.* .v. K; *add.* .v. chapitre HN 183 sensiblement] *add.* que M tout] *om.* D estre] est M estre meu] *om.* D en] EP et occident] *add.* mouvon~ D 184 le] *om.* M journel] *add.* almotus(?) du nuit D aussi] ainsi M .i.] un/ung FHKMNP 184-185 aussi ... ymaginer] *om.* D 185 le] ce K on] l'en F; l'an G parmi] par B 186 comme] *add.* une BCDFGHKMN .ii.$_1$] deus F; deux BHMNP; 2 D .ii.$_2$] deux BHMP; 2 D lesquels] lesquieulx N .ii.$_3$] deux BHMNP; 2 D 187 movent] *add.* point M dequoy] desquelz F; des quielx H li un] lun/l'un BKP un] uns H de] *om.* B 188 apparoist] appert BGH 189 Car] que D ceste] celle M du ciel] *om.* B .vii.] sept BHM; 7 D notable] *om.* DFGM 190 char] chariot BM si] aussi D Ptholomee] Ptolomee F; Tholom^{c} B; Tholomee D 191 a$_1$] *om.* DFG; en B a$_1$ ... nous] le point opposite qui ne nous peult apparoir nous estans deca l'equinoctial M 192-193 le pole] *om.* BK 193 monde] *add.* de H et est] *om.* M; ou et est C 194 senefie] signifie/segnefie DGN par] par/per par H

1 "Char", i.e., "chariot." The image of seven stars (from which "septemtrio/septentrio" or "north" is derived) implies the constellation Ursa Maior, although given the proximity of Ursa Minor to the northern pole, the latter ("The Little Dipper") might be more appropriate. Of the seven stars forming the asterism, also known as "The Plough" or "The Big Dipper" or "The Great Wain/ Wagon", four form the plough itself, or the bowl of the dipper, or the wagon (or chariot). The other three stars are along the shaft(s) of the plough or wagon (or chariot), or the handle of the dipper. In the Arabic image of the constellation, these latter stars are three maidens lined up along the shaft pulling a bier through the heavens.

Ptolemy in his star list seems to prefer to refer to parts of animals, in line with his names of Ursa Maior and Ursa Minor, and not to objects such as a chariot, plough, wain or wagon.

Item: ou milieu de ces .ii. poles dessus dis en la superfice convexe du monde ou de la derreniere espere est a ymaginer un cecle equidistant de ces .ii. poles, c'est a dire que chascune partie ou chascun point de cil cecle est aussi pres de chascun pole comme l'autre partie et comme de l'autre pole. Et selonc ce cecle est fait et mesuré le mouvement journel, et est appellé "le cecle equinocial". Car quant le soleil est la endroit, le jour et la nuit sont equals par tout le monde si comme il apparoistra aprés[2] et c'est appelé "equinoce".

195 ou] en B   ces] ses GN   .ii.] deux BHMNP; 2 D   en] est D; eu E   convexe] *add.* de lepe (?) D   196 ou de] en D   a] *om.* B   un] .i. G   equidistant] qui distant K   ces] ses N   .ii.] deux BHMP; 2 D   197 de cil cecle] *om.* FM   cil] icellui P   198 comme$_1$] *add.* de M   partie ... pole] *om.* M   selonc] secun C   ce] cest F   fait et] *om.* M   199 le cecle] *om.* HN   car] que D   200 si] *om.* D   apparoistra] apparra BDFGHKMN; *add.* en B   201 c'est] est DFGM   c'est ... equinoce] *om.* HN   appelé] *om.* K

2 Cf. Chap. 18, ll. 500-503.

LE .VI.^{E} CHAPISTRE. DU MOUVEMENT DES PLANETES

Combien que le ciel tout ensemble soit meu de jour en jour d'orient en occident aussi comme se la .ixe espere traxist aveques soy les autres esperes de dessous. Toutesvoies chascune des .vii. planetes avecques ce est meue d'autre maniere par son propre mouvement aussi comme au contraire. Mes je me tais a present du propre mouvement de le .viii.^{e} espere.

Et quiconques veult savoir philosophie ou astrologie, il est necessaire que il entende ces .ii. mouvemens lesquiex je vueil declarier premierement par exemple. Car nous voion que une nef puet estre meue en l'yaue en venant devers orient et alant vers occident. Et .i. homme qui sera dedens cele nef puet aler de son propre mouvement, au contraire plus belement, vers orient ou de travers. Et ainsi combien f. 5^{r} que en verité cest hommes s'aprouche touzjours de occident, / toutevoies non pas tant comme si il se seist en la nef. Donques est cest homme meu de double mouvement, dont l'un est le mouvement de la nef et de lui ensemble, et l'autre est son mouvement propre.

Semblablement l'espere de la lune est transportee aveques tout le ciel d'orient en occident par le mouvement journel, mes plus tart et autrement que une estoile fichié, pour ce que ele ha son propre mouvement aussi comme au contraire.

Et pour ce en .i. moys la lune fait un tour moins environ la terre que ne fait une estoile fichié et le solail un tour moins en .i. an. Et les autres planetes en certain temps, chascune selonc la quantité de son propre mouvement.

202 Le ... chapistre] *om.* HKN .vi.^{e}] 6e M planetes] *add.* .vi./.vi.^{e} KN; *add.* .vi. chap~ H 203 ciel] *add.* tourne ensemble et B d'orient] en orient K 204 aussi] ainsi M comme] *rep.* P .ixe] novisme FG; 9me DP; neufiesme M traxist] traixist P; traisist HN 205 des] de F; de ces K .vii.] sept BHM; 7 D maniere] *om.* B 206 propre$_1$] propos K tais] *add.* quant K propre$_2$] propos K 207 .viii.^{e}] 8me P; 8 D; huyt^{e} M; huitiesme N 209 entende] atende D ces] *om.* M; ses N .ii.] deux BHKMNP; 2 D declarier] declairier CKP; declarer FGM; declarier H; descairier N; desclarer B 210 Car] quant D et] en N; *add.* en GK 210-211 devers ... vers] d'orient en M 211 vers] devers K .i.] un/ung BDHKMNP qui ... peut] estant en icelle peult M 212 au ... belement] *om.* FM plus] sont B Et] Car N ainsi] aussi B 213 en] de F en verite] *om.* BN s'aprouche] cy aprouche D 214 si] se GP se seist] ne se mouvoit point M; se asseist K; seist D 214-215 Doncques ... nef] *om.* K 215 dont ... mouvement$_2$] *om.* N est$_2$] *add.* de M 216 propre] *add.* au contraire M 217 Semblablement] Pareillement M tout] *om.* B 219 ele] le K aussi] ainsi M 220 .i.] en F; un/ung BDHKMNP moins] *add.* ou N environ] entour K 220-221 moins ... fichié] *om.* D 221 fichie et le] fichiee pour ce que comme dessus elle a son propre movement aussi comme au contraire et K un tour moins] ymet (?) D .i.] en F; un/ung BDGHMNP les] ainsi des B 222 temps] *add.* une P chascune] *om.* B selonc] secon/secou C

Item: il appiert sensiblement que se vous resgardes la lune a .i. soir emprez une estoile fichié, et resgardes la le soir ensuivant, vous la verres esloignié de celle estoile et sera derrieres, aussi comme plus vers orient, et l'autre soir aprés, encore plus derriere, et ainsi ensuiavment.

Item: posé per ymagination que tout le ciel cessast de son mouvement journal, et que les autres mouvemens demourassent – je di que adonques la lune feroit .i. tour ou .i. circuite aussi comme au contraire en alant de occident en orient, et le feroit en tant de temps comment durent maintenant .xxvii. jours et .viii. heures.[1] Et Mercure et Venus et le solail feroient .i. tour en alant aussi de occident en orient, et en retournant en occident ou l'espace d'un an; et Mars en .ii. ans ou environ; et Jupiter en .xii. ans ou environ; et Saturne en .xxx. ans ou environ. Car maintenant font il leur cours ou leurs revolucions, et acomplissent leurs propres mouvemens es espaces de temps qui j'ai ici nomees.

232 ou$_1$] *read* en

223 que] car FGM .i.] un/ung BDFKMNP emprez une] pres d'une M 224 la$_1$] *om.* N 225 derriere] desrengee et derreniers K aussi] ainsi M comme] *om.* D plus] *om.* M apres] *om.* D; emprez B encore] *om.* HN 226 ensuiavment] en ensuivant P 227 journal] jour el K 228 .i.] un/ung BDFGHKMNP 229 .i.] un/ung BDFGHKMNP aussi] ainsi M au] *rep.* P au contraire] *om.* FM de occident en orient] d'orient en occident B 229-232 et ... an] et en retournant en orient en l'espasse d'un moys D 230 comment] comme GN .xxvii.] 27 MP .viii.] 8 MP 231 et$_1$] *om.* M .i.] un/ung BGHKMNP aussi] ainsi M occident en orient] orient en occident K et$_2$] *om.* M 231-232 en$_1$ ... ou$_1$] aussi comme au contraire en B 232 ou] en BDFGHKMN .ii.] deux PHM; 2 DP et$_2$] *om.* M 233 .xii.] douze B; 12 DMP; .ii. K ou environ$_1$] *om.* B et ... environ$_2$] *om.* K .xxx.] 30 DMP il] *om.* D 233-234 Car ... accomplissent] Et ainsi font ils maintenant leurs revoluucions en accomplissant M 234 leur] leurs BHKNP cours] tours H ou leurs] ou leur FG; et D revolucions] *add.* es espaces de temps cy dessus nommés K et] en M; *add.* aro)e seu et(?) B leurs] leur CF 234-235 es ... nomees] *om.* K 235 de] des H qui j'ai] *om.* HN

1 This is the siderial period (27 days, 7 hours, 43 minutes) measured against the fixed stars.

LE .VIIE CHAPISTRE. DU ZODIAQUE ET DE SES POLES

Le propre mouvement des planetes dessus dit n'est pas tout droit d'occident en orient, ou au contraire du premier mouvement – c'est assavoir du mouvement journel – mais est aussi comme en bieis.

Et ceci puet apparoir sensiblement; car, se vous regardés la lune emprés une estoile fichié, vous la trouveres la seconde nuit, out la tierce, derriere cele estoile non pas tout droit vers orient, mais, aveques ce, sera traité vers septentrion, ou vers midi.

Et pour ce tout aussi comme le mouvement journal est fait et mesuré selonc le
f. 5^{v} cecle equinocial, / si comme il appert ou quint chapitre,[1] ainsi est le propre mouvement des planetes fait selonc un autre cecle qui est appelé "zodiaque".

Et cest cecle ici traverse et interseque l'equinocial en .ii. poins contraires et devisie chascun de ces .ii. cecles, et l'autre en .ii. moitiés par ces .ii. poins. Et l'une moitié du zodiaque decline et se trait vers septentrion, et l'autre vers midi.

Et aussi les poles du zodiaque declinent et sont loing des poles du monde, et sus les .ii. poles du zodiaque sont fais les mouvemens des planetes ou pres.

Cest zodiaque doit estre ymaginé en la .ixe espere ou derreniere aussi comme est l'equinocial, et sont meues les planetes sous cestui cercle.

236 Le ... chapistre] *om.* HKN .viie] 7me MP poles] *add.* .vii. K; *add.* .vii. ca. HN 237-238 d'occident en orient] d'orient en occident N 238 ou] *om* CDFGHKN c'est ... mouvement$_2$] *om.* K 238-239 ou au ... journal] *om.* M 239 aussi] ainsi KM en] es G 240 Et ceci] Cestui B sensiblement] *om.* B empres] apres N; *add. interlin.* de D 241 seconde] 2^{e} D ou] out E ou la tierce] *om.* F; ou la 3^{e} D 242 tout] *om.* B orient] occident K ce] *add.* se BCFGHKN traite] tyres M 243 aussi] ainsi MF selonc] secun C 245 ou] au P quint] .v.^{e} B; 5^{e} D 246 fait] *om.* F; *add.* et mesure B selonc] secun C appele] *add.* le G 247 cest] ce M; de ce N ici] *om.* B .ii.] deux BGHMN; 2 DP 248 de ces .ii.] *rep.* F .ii.$_1$] deux BHKM; 2 DP .ii.$_2$] deux BHMNP; 2 D par ... poins] *om.* B .ii.$_3$] deux HMN; 2 DP 250 aussi] ainsi M declinent et] *om.* M 250-251 Et$_1$ ... pres] *om.* G declinent ... zodiaque] *om.* K 251 .ii.] *om.* HN; deux BM; 2 DP les$_2$] *add.* propre M planetes] sept planetes ou environ iceulx M ou pres] *om.* HN 252 Cest] Ce M ymagine en la] *om.* D .ix.^{e}] 9me DMP; novisme G ou] *om.* K 253 cestui] chascun HN; ce M

1 Above, Chap. 5, ll. 198-201.

LE .VIII.[E] CHAPISTRE. DE LA DIVISION DU ZODIAQUE

Il ha ou zodiaque .iiii. poins principaus, c'est assavoir les .ii. poins dessus dis ou il interseque l'equinocial, et .ii. autres la ou il s'ezloigne le plus et decline [vers] l'equinocial, desqueles l'un est vers septentrion, et l'autre vers midi. Et est le zodiaque devisé en .iiii. parties equales par ces .iiii. poins.

f. 6r / Item: chascun de ces quarties du zodiaque est divisé en .iii. parties equales du zodiaque principauls qui sont appelees "signes".

Ces signes sont nommés ainsi: Aries, Taurus, Gemini, Cancer, Leo, Virgo, Libra, Scorpio, Sagitarius, Capricornus, Aquarius, Pisces. Et n'est mestier de les nommer en francois, car leurs noms ne sont pris de nule propriete qui ait effet quant a la science des jugemens.

Le signe premierement nommé c'est assavoir Aries, se commence ou point la ou le zodiaque interseque l'equinocial, ou quel point le soleil commence[1] a decliner vers septentrion et vers nous, et s'estent cest signe vers orient. Et aprés est Taurus, le secon[t], en venant vers orient, et ainsi des autres ensuiant.

Item: chascun de ces .xii. signes est divisé en .xxx. parties qui sont appelees "degrés". Et ainsi a ou zodiaque .ccclx. degrés.

Item: chascun degré est divisé en .lx. minus, et chascun minu en .lx. secons; et

271 minus] *read* minutes

254 Le ... chapistre] *om.* GHKN .viii.[e]] 8me MP; 8 D zodiaque] *add.* .viii. K; *add* .viii. c./chap~ HN 255 Il ha] Il y a P ou] au P .iiii.] quatre BFHN; 4 DMP c'est assavoir] savoir est (?) B .ii.] deux BHMN; 2 DP 256 .ii.] deux BKMN; 2 DP et_2] *add.* se G vers] *om.* EP; de CDGM; de le F; de vers B 257 l'un] si un F septentrion ... $vers_2$] *om.* D 258 $.iiii._1$] quatre BGM; 4 DP ces] *om.* N $.iiii._2$] quatre KM; 4 DP 259 Item] Et M ces] ses G quarties] 4 D du zodiaque] *om.* B .iii.] trois HM; 3 DP equales] egagales K; *add.* ainsi sont xii. parties BDGHKN 259-260 du ... principauls] et ainsi sont .xii. parties FM 261 Taurus] Thaurus GK 262 Scorpio] Scorpius DP 263 leurs] les K 264 des jugemens] *om.* F 265 premierement] premier M ou] au DP la] *om.* HKN 266 ou quel] auquel P 267 et_1] au P vers nous et] *om.* FM cest] ce M Taurus] Thaurus G 267-268 Et_3 ... orient] *om.* M 268 secont] seconc E en ... orient] *om.* B ensuiant] *om.* D; *add.* apres HN 269 Item] Plus M; *add* de N ces] ses N .xii.] 12 DMP .xxx.] 30 DMP; ~~xxx~~[30] B 270 a] sont B ou] au P .ccclx.] trois cens soixante H; 360 MP; en 360 D; trois cens .lx. N degres] *corr. from* quart D 271 Item] Et M $.lx._{1,2}$] soixante H; 60 MP 271-272 Item ... quars] *om.* D

1 Strictly speaking, the sun begins to move to the north (*septentrion*) at the beginning of Capricorn, i.e., at the winter solstice. Aries begins at the spring equinox. However, Oresme is probably trying to say that at this point (the spring equinox, i.e., the beginning of Aries) the sun moves to the north of the equator/equatorial circle.

chascun secont en .lx. tiers, et chascun tiers en .lx. quars. Et souffist combien que l'en peust proceder en oultre a quins, a sextes, et ainsi tant comme il plairoit.

Semblablement chascun des cecles de l'espere ou quiconques autre soit grant soit petit est ymaginé estre devisé en .ccclx. degrés. Et chascun degré ont [.lx.] minutes et en oultre, comme dit est. Et ainsi chascune heure en .lx. minus, et en oultre, comme dit est.

272 .lx.$_{1,2}$] 60 MP tiers] tierres M Et$_{2}$] *add.* ce M 272-273 que l'en] qu'on P; que l'a G 273 peust] *add.* bien D en$_{2}$] *om.* D quins] quintes et M quins a sextes] minuts et seconds D a$_{2}$] et N tant] *om.* D 274 des] de ces DHK; de ses N de l'espere] *om.* K; des esperes B; et l'espere D quiconques] quelconque M; chascun B autre] *om.* K soit] *om.* B 275 soit] ou DHKMN est ymagine] peut K .ccclx.] .iii^{c}lx. H; 360 DMP; trois cens .lx. N .ccclx. ... ont] .lx. degres et chascun en B 275-276 degres ... .lx.$_{1}$] *om.* F degres ... et$_{1}$] parties et chasune partie M ont .lx.] ont EP; en .lx. DGHN; en .ccclx. K ont ... heure] *om.* D 276 ainsi] aussi FKM .lx.$_{2}$] 60 DMP 276-277 Et$_{2}$ ... est] *om.* N 277 comme] *add.* se B dit est] dessus M; *add.* dessus K

DIAGRAM: CHAPTER 8

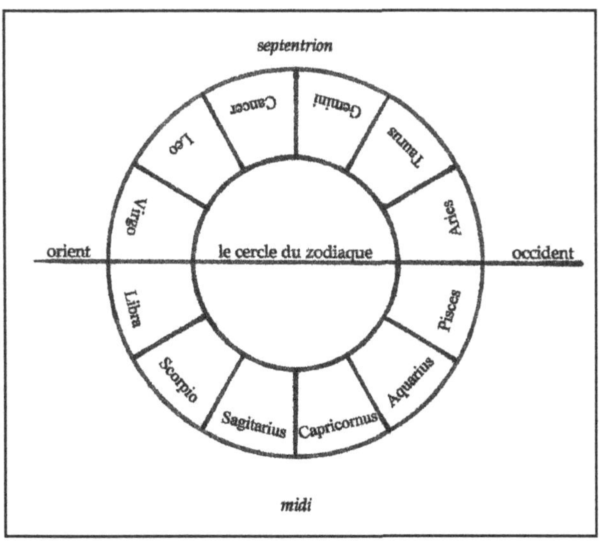

Complete diagram: E BCFGHKNP
Alternative diagram: M
No diagram (or space only): D

septentrion] F orient] *om.* H le ... zodiaque] *om.* BFGHK; mundus N occident] *om.* H midi] F

DIAGRAM: CHAPTER 8 ALTERNATIVE[2]

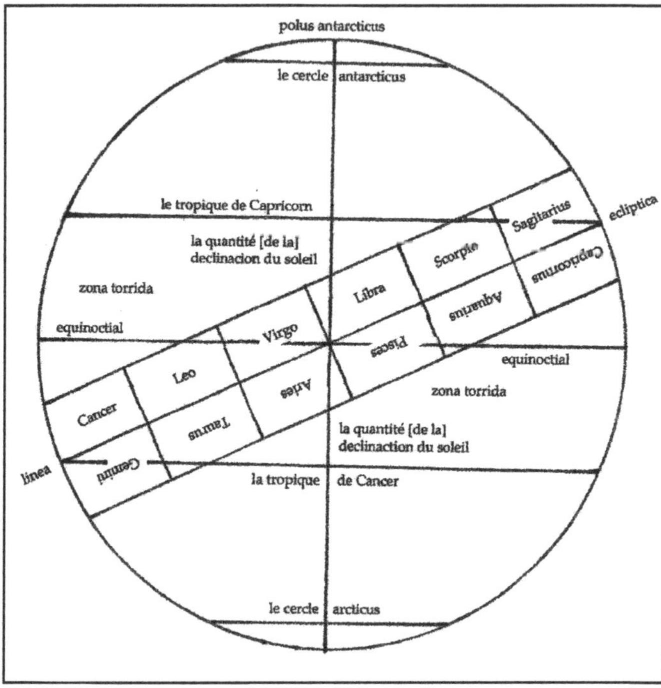

Complete diagram: M

2 This diagram, found only in ms M, is exceptional for being so detailed. Other diagrams tend to be simpler and more "conceptual" in nature. In this one the scribes shows a great deal of technical skill, reminiscent of drawings usually found in more sophisticated works on astronomy or on the astrolabe.

Note that in this diagram north is at the bottom and south is at the top, the opposite of our current practice.

LE .IXE CHAPISTRE. DONT VINT CESTE DIVISION

Chascune espere du ciel est aussi comme d'une meisme nature en chascune de ses parties fors la ou sont les estoiles. Et se il y a aillours variacion, si n'est elle pas distinctee ne cognoissable, car tout est continué et presque semblable, si comme sont les parties de l'air ou de l'iaue. Et pour ce celle division du zodiaque ou d'autre cecle par degrés n'est pas faite ne causee de nature, mais de volenté, selonc raison d'arimetique.[1] Car qui eust voulu, ausi eust l'en dit que chascun signe ha .lx. degrés, et chascun cecle .ccccIxxx. degrés ou en autre nombre, mais le nombre et la division dessus dite est tres couvenable pour cognoistre les mouvemens et l'estat du ciel. Toutesvoies, quant au zodiaque, la premiere division qui est en .iiii. quarts par les .iiii. poins principauls de li est assés fondee sus la nature du mouvement des planetes.

f. 6^{v} Mais que chascun quart soit divisé en trois signes, et chascun / en trint degrés, et fu chose voluntaire, que il n'a plus de difference naturel entre le derrenier degré de Aries et le premier de Taurus que il ha entre le secont et le tiers de Taurus. Et ainsi est il des autres parties du ciel fors tant seulement quant le soleil a aucun regart

278 Le ... chapistre] *om.* GHK Le ... division] De la signe des esperes du ciel N .ixe] 9me DMP division] *add.* .ix. K; *add.* .ix. chap~ H 279 aussi] ainsi M nature] matiere F 279-280 de ses parties] partie HKN 280 il y a ... si] aultre differeance il y a s'il M y a] ma F; n'y a G aillours] nulle HN; aucune K 280-281 si ... cognoissable] *om.* B 281 distinctee] distancee F ne] ou K car ... presque] mais est ainsi comme M presque semblable] cognoissable B semblable] semble F 282 ce celle] telle GH; ce telle K 283 par degres] *om.* B ne] ou K de$_{1}$] par MF selonc] par M; secon/secou C 284 ausi] *om.* B; *add.* en F l'en] il K ha] *om.* F; eust K .lx.] 60 MP 285 cecle] cercles F .ccccIxxx.] .iiii^{c}iiii.^{xx} FGKN; .iiii.^{c} et .iiii.^{xx} B; quatre cens quatre vins H; en 480 M; 480 DP degres] *om.* D ou ... nombre$_{1}$] *om.* HKN mais] neamoin~ D et$_{2}$] qui est selon BDFGHKMN 286 tres] *om.* K et l'estat] *om.* K 287 quant ... est] la division du zodiac dessus dictez M .iiii.] quatre FKM; 4 DP quarts] cas HN; parties M; parties ou quarts P les] *om.* K 288 .iiii.] *om.* G; quatre BFN; 4 DMP li] *om.* M; lui/luy BDFGHKNP 290 soit] fust K en trois signes] *om.* M trois] .iii. FGKN; 3 DP et chascun] *om.* B; *add.* signes CDFGHKMN en trint degres] *om.* M trint] .xxx. HKN; tant? E; *add.* de DFG 291 et] ce FGHKN que] car FGHMN que ... plus] car il n'y a rien D plus] point M naturel] *om.* M 292 de Taurus$_{1}$] *om.* N; de Thaurus G que il ha] que il n'a K; que il y a P; car il y a D; ne M secont] 2 D tiers] 3 D de Taurus$_{2}$] Thaurus G 293 quant] que HKN quant le soleil] *om.* M

1 Oresme is saying that the division into 60ths (sexagesimal measure) is not "from nature" but is an arbitrary system set up by humans for reason of ease of calculation. In his example he notes one could arbitrarily divide a sign into 60 "degrees" (rather than 30) and a circle into 480 "degrees" (rather than 360). Note that there would then only be 8 signs of 60 degrees.

a une estoile fichiee. Il a aucune signification autre ou a une autre que quant il est autrement. Mes elles ne sont pas tousjours en .i. meisme degré ne en .i. signe.

Et pour ce pluseurs regles des jugemens d'astrologie sont faulses qui furent escriptez premierement par aucuns qui ne savoient pas philosophie naturel ou qui n'avoient pas congneu ou apperceu le mouvement des estoiles fichiés, ou qui estoient trop hardis de sentencier de quoy l'en voit et a veu communement que ceuls qui presument a jugier par teles regles faillent a voir dire.

294 une] aucune BFK fichiee] *add.* ou a une autre BDFGHMN Il ... autre$_1$] *om.* K ou a une autre] *om.* DFGHMN 295 .i.$_1$] un/ung BDFHKMNP ne en .i. signe] *om.* H .i.$_2$] un/ung MKNP signe] meisme signe KN 296 des] de BH 298 congneu ou] *om.* HKN ou$_1$] ne M le mouvement] les mouveements FHN fichies] *om.* D 299 de quoy] dont M l'en voit] le voit on B et a vue] *om.* M a] *add.* les BD 300 a$_1$] *om.* K teles] icelles M voir dire] dire verité M

Le .x.^{e} chapistre. De la latitude du zodiaque

Chascun cecle de l'espere est ymaginé comme une grelle line ou comme un tenue filet sanz laise, fors le zodiaque tant seulement qui est comme .i. large chapiau. Et aussi comme il a en son tour .ccclx. degrés, il a de largesce ou de latitude .xii. degrés ou environ. Et ainsi chascun signe a .xxx. degrés de long et .xii. de lé.

Et est le soleil meu et ha son cours et sa voie parmi le milieu de ce cecle selonc un grelle cecle qui est apellé "la line eclipcique".

Et la cause pourquoy le zodiaque est ainsi large, c'est pour ce que les autres planetes n'ont pas leur cours ne leur voie sous la line eclipcique parmi le milieu du zodiaque aussi comme ha le soleil, mes declinent et se traient plus vers midi d'une part, et vers septemtrion d'autre.

De quoy il avient et le puet ou veoir sensiblement que la lune vient aucune fois plus haut vers nos testes que ne fist onquez le soleil, et aussi descent elle plus bas et se trait plus vers midi, et ainsi est il des autres planetes.

Car aussi comme le zodiaque decline de l'equinocial si comme fu dit ou .viie chapistre,[1] semblablement, mais non pas tant, declinent les cecles des planetes de la voie du soleil, qui est appellee "eclipcique" pour ce qu'il ne puet estre eclipcé de soleil ne de lune fors quant la lune est en celle voie, ou bien pres, si comme il apparoistra aprés plus plenierement.[2]

301 Le ... chapistre] *om.* GHKN .x.^{e}] 10me DMP De ... zodiaque] *om.* B; *add.* .x. K; *add.* .x.^{e} c. N; *add.* .x. chap~ H du] au D 302 un] .i. G tenue] muun(?)P; tenure FH 302-303 grelle ... laise] ligne sans largeur M ou ... zodiaque] *om.* B 303 est] *add.* ainsi M .i.] un/ung BDFGHKMNP large] *om.* HKN 303-304 Et ... il$_{2}$] qui M 304 .ccclx.] .iii^{c}lx. H; 360 DMP; trois cens .lx. N il$_{2}$ ... .xii.] et de large 12 M .xii.] 12 DP 305 ainsi] aussi BFHNP .xxx.] 30 DMP .xii.] 12 DMP lé] large M 306 parm i] par M selonc] secun C 306-307 de ... cecle] d'icellui zodizc en ung cercle grisle M 307 un] .i. G la line] *om.* BM 308 c'est] est D 309 leur$_{1}$] leurs GHMNP ne leur voie] *om.* K leur$_{2}$] leurs M parmi] par M 310 aussi] ainsi M ha] *om.* GM mes] *add.* se CG traient] retrayent vers] *add.* le B 312 ou] l'en F; l'ou B 313 vers] *add.* nous et sus K 314 plus] *om.* B est il] *om.* D 315 aussi] ainsi M decline] *om.* G comme$_{2}$] *add.* il BDFHK fu] il a esté M .viie] 7me MP; 7 D 316 de] du G 380 qu'il] car il K 318 fors] *add.* tant seulement G celle] ceste F; tele H 319 apparoistra] apparra BDGHKMN; apparru F apres] *om.* M plenierement] plainement M

1 Chap. 7, ll. 247-249.

2 Chap. 46, ll. 1453-1459 and following.

LE .XIE CHAPISTRE. COMME SIGNE PUET ESTRE PRIS EN TROIS MANIERES

Signe en sa propre significacion est une superficie ou une partie du zodique en la f. 7r derreni/ere espere, laquele partie ha .xxx. degrés de long et .xii. degrés de lé, comme dit est.[1] Et ainsi quant on dit que une planete est en Aries ou en .i. autre signe, c'est a dire que elle est sous Aries ou sous tel signe.

En la seconde signification, l'en puet entendre par .i. signe tout ce que est sous celle superficie ou partie ou signe pris en la premiere signification. Et ainsi un signe est aussi comme un coing ou comme une cheville acue de quoy la pointe est ou centre du monde et le gros bout en la derreniere espere. Et cil coing ha .iiii. costés desquels les .ii. sont plus larges que les .ii. autres. Et en ceste maniere est le soleil ou une autre planete tousjours en aucun signe et non pas sous aucun signe.

Tiercement on puet ymaginer .vi. cecles passans par les commencemens des .xii. signes et par les poles du monde.[2] Et aussi comme .i. cecle divise l'espere en .ii. parties, .ii. cecles en .iiii. en la guise que une pomme seroit devisee en .iiii. quartiers, semblablement ces .vi. cecles divisent toute l'espere du monde en .xii. parties equales qui sont appellees ".xii. signes". Et qui entent signe en ceste maniere

320 Le ... chapistre] *om.* HKN; Le .xii. c. F Le ... manieres] *om.* G .xie] 11/11me DMP Comme] Commant DM; Comment H signe] figure FN trois] .iii. FHN; 3 D; .iiii. K; plusiers M manieres] *add.* .xi./.xie KN; *add.* .xi. ca. H 321 Signe] *add.* est H zodiaque] *add.* qui est B 322 .xxx.] 30 DMP .xii.] 12 DMP degres$_{2}$] *om.* BD lé] l'arc F; large M comme] si comme N 323 ainsi] aussi BP quant] que D que] *om.* P .i.] *om.* FGHMNP; un/ung BDK autre] aucun FG 323-324 c'est ... signe] convient entendre qu'elle soit soubz telle superficie ou a l'endroit d'icelle M 324 a dire] comme G 325 En$_{1}$ ... signe] *om.* D l'en] *interlin.* C .i.] un/ung/une BFHKNP 325-326 En$_{1}$ ... significacion] *om.* M 326 celle] telle P ou$_{2}$] en HN un] *om.* M; .i. G 326-327 Et ... acue] *om.* D 327 aussi] *om.* M un] .i. G comme$_{2}$] *om.* BG ou$_{2}$] au P 328 bout] *add.* est B cil] cil qui C; icellui P .iiii.] 4 DMP 329 .ii.$_{1}$] 2 DMP; deux BHN .ii.$_{2}$] *om.* D; 2 MP; deux BHN ou] en HN 329-330 le ... tousjours] tousjours le soleil en aulcune planete M 330 une] *om.* BK aucun$_{1}$] autre H; un autre N aucun$_{2}$] autre HN 331 Tiercement] Apres M .vi.] 6 DP; six M 332 .xii.] 12 DMP aussi] ainsi M comme] *om.* D .i.] un/ung PDFGHKMN; le B cecle] *interlin.* D .ii.] *om.* B; 2 DP; deux HKMN 333 .ii.] 2 DP; deux BM .iiii.$_{1}$] 4 DMP; quatre BFHK en$_{2}$ ... .iiii.$_{2}$] *om.* K en$_{2}$ ... quartiers] *om.* D .iiii.$_{2}$] 4 P; quatre FM 334 .vi.] 6 P; six M toute] *om.* HKN du monde] *om.* M .xii.] 12 DMP 335 .xii.] 12 DMP; douze H

1 Chap. 10, l. 305.

2 That is, a spherical wedge or ungula. Technically these lines should intersect at the poles of the ecliptic or zodiac, rather than the poles of the earth, but Oresme has clearly chosen the poles of the earth.

chascune estoile et chascune partie du ciel et de la terre est en aucun des .xii. signes combien que elle soit loing du zodiaque.

Toutesvoies de ces trois significations, la premiere est la plus propre selonc laquelle les signes sont parties du zodiaque.

Ainsi donques ai je dit des .ii. plus principauls cecles de l'espere desquels l'equinocial est aussi comme la droite et grellete ceinture du monde ou du ciel, et le zodiaque est comme une large renge qui traverse en biés et qui est aussi comme ferree gentilement ou figuree et noblement paree des ymages des signes entailliés subtilment et composés de estoiles fichiee aussi comme d'escarboucles luisans ou de precieuses gemmes plaines de grans vertus et par mestrise assises tres excellentement.

336 en] *om.* G aucun] autre HN des .xii. signes] signe M .xii.] 12 DP 337 que ... soit] quelles soient HN 338 trois] *om.* M; .iii. BFGLN; 3 D la$_1$] ici la M selonc] secun C 340 .ii.] 2 DP; deux BHN 340-341 ai ...ceinture] je dys des deux principaulx cercles dont l'equinoctial en est l'ung lequel est ymaginé comme une gresle sainture M 341 et$_1$] *add.* la N ou du ciel] *om.* HKN le] *illeg.* B 342 est$_1$] *om.* G; *add.* aussi BHKN renge] sainture M en bies] au *illeg.* B; en byais l'equinoctial M aussi] ainsi M 343 gentilement] rome/roment D signes] figures K 344 de] des BM aussi] ainsi M 345 precieuses ... plaines] pierres precieuses gemmes plaines M 346 excellentement] noblement HKN

LE .XII.[E] CHAPISTRE. DES .II. COLURES

En l'espere ou ou ciel sont ymaginés .ii. cecles appelés "colures", desquels li uns passe par les poles du monde et par les poles du zodiaque et par les .ii. poins des .ii. solstices, c'est assavoir par le premier point de Cancer et par le premier point de Capricornus. Et l'autre colure passe par les poles du monde et par les .ii. poins equinociauls; c'est assavoir par le premier point d'Aries et par le premier point de Libra.

f. 7v Ces .ii. colures s'entrefforchent ou intersequent / en crois en chascun des poles du monde, et divisent toute l'espere du monde en .iiii. quartiers equauls.

Par le colure qui passe par les solstices est mesuree la declination du zodiaque et de ses poles. Car la plus grant declination du zodiaque est en ces .ii. poins desquels le point du solstice d'esté, c'est assavoir [le] premier point de Cancer, decline vers septentrion ou est loing de l'equinocial par .xxiii. degrés et petit plus. Et ses degrés sont pris et signéz en cest colure et autant, ne plus ne mains, decline de l'autre part; et est loing de l'equinocial le point du solstice d'yver qui est le premier point de Capricornus.

De quoy il s'ensuit par necessité que autant declinent et soient loing les poles du zodique des poles du monde.

347 Le ... chapistre] *om.* KN; Le .xii. c. F   Le ... colures] *om.* G; De deux colurres. .xii. chap. H   .xii.[e]] 12/12me DMP   Des] De FHKN   .ii.] 2 D; deux BFKMP   colures] coullinas P; *add.* .xii. K; *add.* xii[e] c. N   348 ou ou] ou B; ou au P   .ii.] 2 DP; deux HMN   349 poles$_2$ ... par$_3$] *om.* B   .ii.$_1$] les .ii. FG; les 2 DP; deux BH; les deux M   .ii.$_2$] deux BDMP   349-350 des .ii. solstices] *om.* H; de solstice HN   350 point$_2$] *om.* DHKN   351 les$_1$] *add.* deux M   .ii.] *om.* N; 2 P; deux BHM   poins] *om.* M   351-352 les$_1$ ...par$_1$] *om.* D   352 point$_1$] *om.* DB   point$_2$] *om.* BDM   354 .ii.] *ms* A *begins*; 2 D; deux BHMN   ou] *om.* HKN; et M   intersequent] *om.* HKN; sont reforchees B; introsequent MP   crois] .ii. B   355 et ... monde$_2$] *om.* K   .iiii.] 4 P; quatre ABFHKM; 4 quatre D   quartiers] pars/parties D   356 declinacion] *add.* du soleil ou M   357 et ... zodiaque] *om.* N   .ii.] 2 DP; deux MN; trois B   358 le] *om.* EP   359 ou] qui DP   est loing] il est esloigné M   .xxiii.] 23 DP; .xxxiii. K   et petit] ung petit M   360 Et$_1$] *add.* de FM   ses] ces CFHM   pris et] *om.* D   et$_2$] es H   361 de$_1$] *om.* P; il de K   et est eloing] *om.* D; et s'esloingne K   361-362 le$_2$ ... de] *om.* HKN

LE .XIIIE CHAPISTRE. DU MERIDIAN

Un autre cecle est en l'espere materiel et ymaginé ou ciel qui est appellé "meridian", et passe par les poles du monde et par le point du ciel qui est tout droit sus nostre teste, lequel point est appelé "cenith". Et toutesfois que le soleil est venu d'orient jusques a ce cecle, il est midi – et pour ce est il appellé "meridian" – et est la moitié de ce cecle sus terre et l'autre dessous, qui passe par le point de mienuit qui est opposite a nostre cenith.

Et toutefois que un homme se trait ou est meu vers orient ou vers occident, il ha autre meridian nouvel. Et pour ce est il plus tost midi a ceuls qui sont plus vers orient, et plus tart a ceulz qui sont ou regart des autres plus vers occident.

Et se un homme ne se meut d'un lieu, son meridian ne se mue, mes est touzjours tout .i. meisme. Et se il va tout droit vers septentrion ou vers midi, sans decliner a orient ou a occident, encore a il touzjours .i. meisme meridian, mes toutesvoies il ne se puet remuer que il n'ait continuelment autre cenith.

365 Le ... chapistre] *om.* KHN   Le ... meridian] *om.* G   .xiii.^{e}] .xiii. F; 13/13me DP   meridian] *add.* .xiii. KN; *add.* .xiii. chap. H; *add.* grand cercle de la sphere D(*later hand*)   366 et] *om.* HKN; est et est B   ou] au P   367 tout] *om.* K   368 point$_2$] *om.* D   toutesfois] toutes les fois M   369 jusques] siques ABCFGH   ce$_1$] celui M   est$_2$] *om.* H   370 ce] se G   sus] dessus M   371 a] de N   372 un] .i. G   se ... meu] se trait B; est meu ou se trayt M   373 plus$_2$] *om.* DGM   374 a ceulz] *om.* B   qui sont] *om.* BFMP   ou] au MP   autres] *add.* qui sont BFM   plus$_2$] *om.* MN   375 meridian] *add.* aussi M   mue] *add.* point M   376 tout$_1$] *om.* G; soubz M   .i. mesme] ung mesmes meridien M   .i.] un/ung ABDFGHKNP   376-377 Et ... meridian] *om.* M   377 encore a il] il ha B   .i.] un/ung ABDFHKNP   378 continuelment] *om.* B; tousjours K; *add.* une G   autre] *rep.* F

Le .xiiiie chaspitre. De l'ozizon

Encore ha .i. autre cecle en l'espere qui est appellé "ozizon", lequel devise la partie du ciel que nous voions sus nous d'avequez celle que nous ne voions pas, et qui est aussi comme sous nous.

Et pour ce que la terre n'est fors tant seulement que aussi comme .i. petit point ou regart du ciel, il est ainsi que toutefois que un homme est en plan[e] pais, il puet f. 8r veoir la motié du ciel, et ceste / moitié est appellee "nostre hemispere", c'est a dire "nostre demie espere". Et ainsi cil cecle qui est appellé "ozizon" devise le ciel en .ii. moitiéz. Et cecle ici et le meridian s'entre intersequent et divisent en croiz chascun en .ii. moitiéz. Et le point du meridian que est sus nous appellé "cenith" est pole de l'ozizon. Et chascune persone est el[le] centre dessouz l'ozizon. Et pour ce, ainsi comme toutesfois que .i. homme se transporte de lieu en autre il est en autre endroit du ciel et ha autre cenith.

Aussi a il autre ozizon nouvel quel part que il voise vers orient ou vers occident, vers midi ou autre part.

379 Le ... chapistre] *om.* K; .xiiii.^e c. N; Le .xiiii. c. F Le ... l'ozizon] *om.* G; De l'orizon .xiiii. chap. H; *add.* .xiiii. K .xiiii.^e] 14/14me DMP De] est de M 380 ha ... l'espere] en l'espere a ung aultre circle M .i.] un/ung ABDFHKNP en] a K 381 d'avequez] de BDHKN celle] telle P et] *om.* HN 382 aussi] ainsi M 383 n'est] si N fors] *om.* K tant seulement] *om.* M que$_2$] *om.* D que aussi] *om.* HN aussi comme] *om.* BK .i.] un/ung ABDFHKNP 384 ou] au P plane] plani E; plat HN 384-385 il$_1$ ... veoir] toutefois nous sommes sus terre comme si ellle estoit plane et ne povons veoir que M 385 nostre] *om.* F c'est] et est B 386 ainsi] *om.* K cil] cellui M; icellui P .ii.] 2 P; deux BHM 386-387 qui ... cecle] *om.* D 387 s'entre] s'e M; s'en HN; se en G crois] trois HN; .iiii. C'est assavoir K chascun] *om.* M 388 .ii.] 2 D; deux HMNP nous] *add.* est K cenith] ceruth F[1] pole] le pole M; appellé le pole D 388-389 de l'ozizon] d'orizon F 389 Et$_1$ ... l'ozizon$_2$] *om.* P; *marg.* E; *add. in marg.* Et chascune prosonne est elle centre des[*illeg.*] orizon P elle] el E; ou/on ABCDFGHKMN dessouz] de son BDFGHKMN; *corr. to* de son AC l'ozizon$_2$] orizon AGN 389-390 pour ... toutesfois] ainsi B ainsi ... toutesfois] toutes et quantesfois M 390 comme] *om.* N .i.] un/ung ABDFHKMNP autre$_1$] *add.* a donc M 390-391 de ... ciel] d'un endroit a autre B 391 et ha] et F; et en K;il ha B; parquoy il a M cenith] ceruth F 392 Aussi a il] et BM quel] quelque MN voise] *om.* F; aille sont B vers$_1$] *add.* midi ou B vers$_2$] *add.* septentrion ou B occident] *add.* ou AFGKM 393 vers ... part] *om.* B

1 Here, and elsewhere in ms F, the writing of "ni" in "cenith" (to this point a connected triplet), is separated into a distinct "r" and "u". This is more an calligraphic issue than an orthographic one.

LE .XVE CHASPITRE. DES .II. MANIERES DE OZIZONS

Combien que en chascun lieu de la terre ait tousjours autre ozizon, toutesvoies tout ozizon est droit ou oblique. Ceuls ont droit ozizon qui habitent dessous l'equinocial,[1] qui ont leur cenith en l'equinocial, car leur ozizon interseque et divise l'equinoctial en croiz et a dr[oit] angles. Et sont les .ii. poles du monde en leur ozizon ainsi que nul de ces poles n'est eslevé dessus leur ozizon ne deprimé dessous.

Mais les autres gens qui habitent allours que sous l'equinocial ont ozizon oblique, car leur ozizon interseque et divise l'equinocial de byais, non pas droit en crois, mes a angles obliques et qui ne sont pas equals. Et leur appert tous temps .i. des poles du monde eslevé dessus leur ozizon.

Et l'autre leur est tousjours mucié et deprimé de sous lor ozizon plus ou mains selonc diverses habitacions, si comme il apparoistra aprés plus a plain.[2] Et pour ce, a il en l'espere materiel .ii. cecles dont li un represente le droit ozizon et l'autre, oblique; et le puet on mouvoir et faire oblique plus ou mains selonc diverses habitacions.

394 Le ... chapistre] *om.* HK; .xve c. N; Le .xv. c. F Le ... ozizons] *om.* G .xve] 15/15me DMP Des] de N .ii.] deux BFHMP; 2 D ozizons] *add.* .xv. K; *add.* .xv. chappitre H 395 tout] *om.* P 397 qui ... l'equinocial$_1$] *om.* M leur$_1$] la D cenith] ceruth F en] dessoubz D car ... l'equinoctial$_2$] *om.* K 398 a droit] adres CE; a .ii. G; en droiz K .ii.] *om.* N; 2 P; deux BM ozizon] *add.* en tel/telle manniere ABDFGHKMN ainsi] *om.* ABDHKMN 399 ces] ses G 400 ont] ou F ozizon] l'orizon G 401 de byais] *om.* G; en biez K 402 tous temps] tousjours M .i.] un/ung ABDFHKMNP 404 mucié] occulte M et$_2$ ... mains] *om.* K 405 selonc] secon/secou C apparoistra] apperra ABDFGHKMN; apporestra P 406 a il] *om.* M materiel] *add.* a M .ii.] deux BHMNP; 2 D 407 oblique$_1$] l'oblique ACDFGHKN on] en A; l'en F mouvoir] mover G oblique$_2$] obliquer M selonc] secun C diverses] *add.* nactions et N

1 We would say "on the equator" rather than "beneath/below the equatorial circle" which to us might also mean "south of the equator."

2 Chap. 20, ll. 547-578; Chap. 21, ll. 581-634.

LE .XVI.E CHAPISTRE. DE L'ELEVACION DU POLE

Comme il appert par le chapistre precedent, nul des poles du monde n'est eslevé sus l'ozizon de ceulz qui habitent sous l'equinocial et qui ont l'espere droite. Mes partout ailleurs .i. des poles est eslevé sus l'ozizon oblique, et tant est le pole plus eslevé et l'ozizon est plus oblique, et tant est l'ozizon plus oblique et le pole est plus eslevé; et de tant est l'autre pole plus deprimé sous l'ozizon.

Et est assevoir que autant comme le pole est eslevé sus l'ozizon, autant a il de distance entre cenith et l'equinocial. Ceste regle appert par ce: car de l'ozizon f. 8v jusques a cenith, qui est pole de l'ozizon, la distan/ce e[s]t par la quarte partie du meridian.

Item: du pole du monde jusques a l'equinocial aussi a il distance par la quarte partie du meridian. Et ces .ii. quars sont equals; donques, qui osteroit de chascun autant comme de l'autre, les demourans seroient equals.

Or oston donques de chascun de ces .ii. quars ce qui est entre le pole et cenith, et ainsi seront equals la partie du meridian qui demeure entre l'ozizon et le pole, qui est appellee "l'elevacion du pole", et la partie d'entre le cenith et l'equinocial.

Item: je le monstre par example en l'ozizon de Paris la ou le pole est eslevé aussi comme par .xlix. degrés.[1] Je di donques que il a entre le cenith de Paris et l'equinocial .xlix. degrés, et le preuve; car de l'ozizon jusques au cenith a .xc.

409 Le ... chapistre] *om.* HKN; Le .xvi. c. F .xvi.e] 16/16me DMP pole] *add.* .xvi. K; *add.* .xvi. ca. HN 411 sous] sessoubs M 412 .i.] un/ung ABDFHKMNP poles] *add.* du monde K sus] et N l'ozizon] *add.* est plus N 412-414 sus ... est] et tant plus est esleve tant plus l'orizon est oblique et M 413 tant ... et$_2$] *om.* H 413-414 et$_3$ ... l'ozizon] *om.* N 414 de] *om.* F pole] *add.* est M sous] dessoubz H 415 assevoir] asses K autant] tant FM comme] que M de] *om.* ADFGHM 416 entre ... et] d'autant est loing le zenit de M cenith et l'equinocial] ceruth et equinocial F par ce] *om.* M 417 jusques] siques ACFG jusques ... l'ozizon] *interlin.* D a] au M est$_1$] est le M est$_2$] *om.* F; et ECP 419 Item] Et M jusques] siques ACFG a$_2$ ... par] la distance si est de M quarte] 4 D 419-420 Item ... meridian] *om* N 420 .ii.] 2 DP; deux BHM osteroit] est D 422 Or] Car N de chascun] *om.* M; *add.* orizon G .ii.] 2 D; 2 ung P; deux BHM et] et le BM cenith] ceruth F 423 la] ca K 424 l'elevacion] la levacion P d'entre le] *om.* N le] *om.* BK cenith] ceruth F 425 Item] *om.* B je] *om.* H monstre] *add.* aussi M 426 aussi] *om.* M .xlix.] *illeg.* D; 49 BMP cenith] ceruth F 426-427 degres ... .xlix.] *om.* DG 427 .xlix.] 49 BMP jusques] siques ACFG au] a HKN .xc.] .xc.^{90} P; nonante ADFGHKN; 90 BM 427-428 et ... degres] *om.* B

1 Oresme's figure is fairly accurate. Actual latitude is $48^{\circ} 51'24''$ but the difference of 8-9 minutes would be hard to measure.

degrés – c'est le quart du cecle. Or en rabatés l'elevacion du pole, si demeure .xli. degrés entre le pole et cenith. Et il a du pole jusques a l'equinocial .xc. degrés; donques rabates en ces .xli. degrés qui sont entre le pole et le cenith, il demeure .xlix. degrés entre le cenith et l'equinocial. Item: il a du cenith jusques a l'orison par l'equinocial .xc. degrés desquels, ostee la distance de cenith jusques a l'equinocial, qui est .xlix. degrés, il demeure .xli. degré[s], c'est l'elevacion de l'equinocial sus l'ozizon de Paris, qui est equale par necessité a la distance d'entre le pole et cenith.

Item: le point du solstice d'esté decline de l'equinocial vers septentrion et vers notre cenith par .xxiii. degrés ou environ, si comme il fu dit ou .xiie chapistre;[2] lesquels degrés adjoustés aveques l'elevacion de l'equinocial, ce seront .lxiiii. degrés. Et ce est l'elevacion du solstice de esté sus l'ozizon de Paris.

Illeques est le soleil a midi quant il est au plus hault vers nostre cenith a Paris.

Item: le solstice d'yver decline de l'autre part vers midi semblablement par .xxiii. degrés lesquels rabatus de l'elevacion de l'equinocial, il demeure .xviii. degrés.

Ce est l'elevacion du solstice d'yver a Paris quant le soleil court au plus bas. Et ainsi qui scet une de ces elevacions, il puet legierement trouver les autres. Et semblablement puet l'en faire en chascune region.

428 c'est] ce pour B quart] 41 D du$_{1}$] d'un B l'elevacion] la levacion P si demeure] de demourra M .xli.] 41 DMP 429 degrés$_{1}$] *om.* B; *add.* qui sont M et$_{1}$] et le B a] y a MP jusques] siques AFG .xc.] .xc.^{90} P; 90 DM; nonante ABFGHKN 430 ces .xli. degres] 41 D .xli.] 41 MP qui ...demeure] il reste M et le cenith] *om.* ABDFG le] *om.* K 431 .xlix.] 49 DMP entre ... et] qui est la distance du zenith jusques a M Item] Plus M a$_{1}$] y a MKP jusques] siques ACBFG 432 .xc.] .xc.^{90} P; 90 DM; nonante ABFGHKN de] du M jusques] *om.* M; siques ABCFG jusques a l'equinocial] *om.* D 433 .xlix.] 49 DMP il demeure] et demeurent M .xli.] 41 DMP degres$_{2}$] degre EABCFGHKN l'elevacion] la levacion P 434 et] et le M 435 de] *om.* P; a B 436 .xxiii.] 23 DMP .xii.^{e}] 12/12me DMP 437 lesquels] les D; lesquieulx N .lxiiii.] 64 BDMP 439 au] le N 441 .xxiii.] 23 DMP lesquels] desquels K; lesquieulx N de$_{1}$] *om.* K .xviii.] 18 DMP 441-442 de$_{1}$ ... degres] *om.* G 443 le soleil] *om.* K 444 scet] seroit HKN puet] pourroit/pouvroit N legierement] legie|legierement N 445 l'en] le F region] *add.* Si comme il peut apparoir par ceste presente figure qui icy apres est figuree G

2 Chap. 12, l. 359.

DIAGRAM: CHAPTER 16

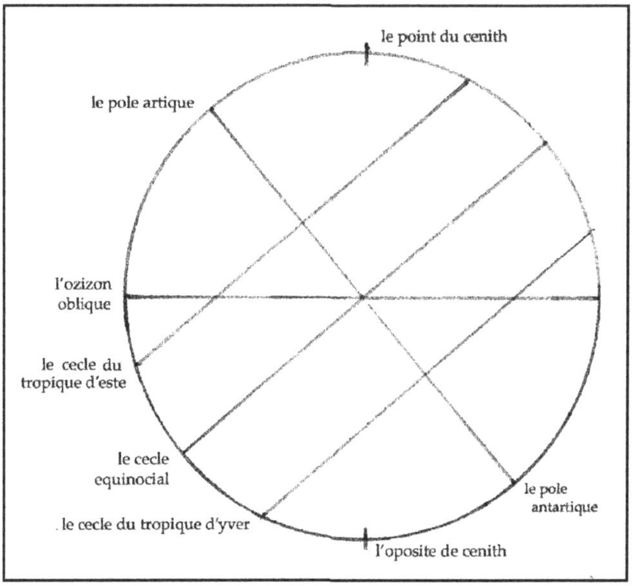

Complete diagram: E CP
Alternative diagram: BFGHKMN
No diagram (or space only): D

DIAGRAM: CHAPTER 16 ALTERNATIVE

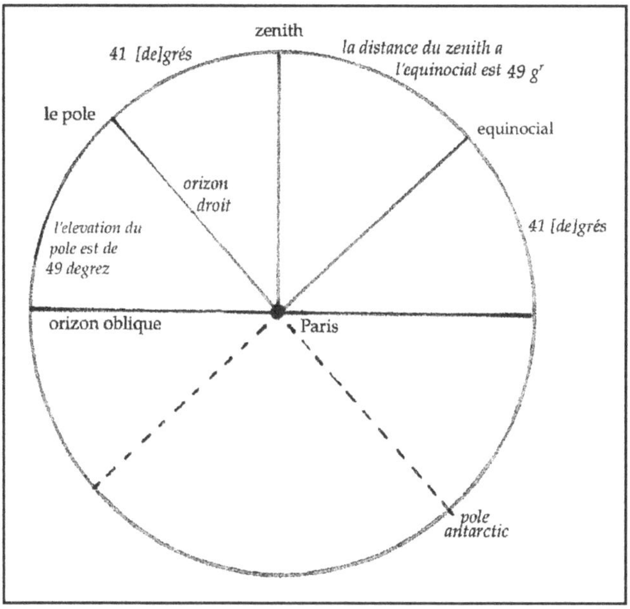

Complete diagram: BFGHKMN

41 [de]grés] M   la distance ... 49 g^r] M   le pole] le polle B; pole arctic M   orizon droit] M   l'elevation ... dégres] M   41 [de]grés] M   Paris] *om.* G; terra M   pole antarctic] M

f. 9r /

LE .XVII.E CHAPISTRE. DES .IIII. MENDRES CECLES

J'ai dit devant de .vi. cecles, c'est assavoir de l'equinocial, du zodiaque, de .ii. colures, du meridian, et de l'ozizon,[1] auquel ozizon il respont .ii. cecles en l'espere materiel. Et chascun de ces cecles est appellé "grant", pour ce que chascun divise l'espere en .ii. moitiés.

Mais encore a il en l'espere .iiii. autres cecles qui sont diz "mendres" ou petis pour ce que chascun d'iceuls divise l'espere en .ii. parties non equauls. Et sont mendres que les autres. Et nientmoins, chascun a .ccclx. degrés aussi bien comme le plus grant.

Les .ii. poins de ces .ii. solstices par leur mouvement journel descripsent .ii. des cecles dessus dis qui sont appellés "tropiques", c'est a dire "conversis".[2]

Car quant le soleil i est, il se convertist a retorner vers l'equinocial. Celui de ces .ii. cecles qui passse par le solstice d'esté – c'est par le premier point de Cancer – et est appellé le "tropique d'esté" et l'autre est appellé le "tropique d'yver".

446 mendres] *read* moindres ("lesser")

446 Le ... chapistre] *om.* HK; Le .xvii. c. F; *add.* est M Le ... cecles] *om.* G; c. *marg.* N *and add in bottom marg.* Des .iiii. mendres cercles .xvii.e c. .xvii.e] 17/17me DMP Des] De K .iiii.] *om.* D; 4 P; quatre BMF mendres] meridiens B cecles] *add.* .xvii. K; *add.* .xvii. chapitre H 447 dit] *add.* ex P de$_1$] des M .vi.] 6 DP; six M de$_3$] des BN .ii.] 2 D; deux BHMP 448 et de l'ozizon] *om.* K l'ozizon$_1$] orizon AFGH respont] corespond M .ii.] et D; 2 P; deux BHMN 449 ces] *add.* .ii. K; *add.* deux B chascun$_2$] *add.* d'eulx K 450 .ii.] 2 DP; deux BHMN 451 .iiii.] 4 DP; quatre BKM mendres] mudiens *corr. to* merindres(?) B mendres ou petis] ou petis ou mendres N 451-452 Mais ... parties] *om.* H 452 pour] pour|pour N .ii.] 2 DP; deux BMN non equauls] inegales M 453 mendres] masdre B a] est divisé par entendement en M .ccclx.] .iiic lx. BH; 360 DMP; trois cens .lx. N 455 deux$_{1,2}$] 2 DP; deux BHKM de ces] des BFMP .ii.$_3$] 2 DP; deus F; deux BHMN des] d'iceulx M 456 dessus dis] *om.* M qui sont] *om.* G 457 Car ... retorner] et retornans M 458 .ii.] 2 DP; deux BHM et] *om.* ABDFGHKM 459 le$_1$] *om.* D d'este ... tropique$_2$] *om.* HN l'autre] *add.* qui passe par le premier point de Capricornus M est appelé] on appelle D le$_2$] *om.* DM

---

1 Equator: Chap. 5, l. 198; zodiac: Chap. 7, l. 246; colures: Chap. 12, l. 348; meridian: Chap. 13, l. 366; and horizon: Chap. 14, l. 380.

2 "Conversis" because this is the line at which the sun "turns around" and heads south again (Tropic of Cancer) or north again (Tropic of Capricorn). See the next sentence.

Les .ii. autres petis cecles sont descrips par le movement journel des poles du zodique entour les poles du monde. Et est l'un appellé le "cecle artique", qui est tout sus nous entour le pole artique.

Et l'autre, qui est tout sous nostre ozizon vers le pole antartique,[3] est appelé le f. 9v "cecle antartique". Ces / .iiii. cecles sont pareillies ou equidistans, c'est a dire que de chascun a chascun a telle distance en .i. endroit ou en une partie comme en l'autre; et n'est pas aussi du zodiaque ou regart des autres cecles.

Item: la distance du pole artique au cecle artique et celle du tropique d'esté jusques a l'equinocial, et celle de l'equinocial jusques au tropique d'yver, et celle du cecle antartique jusques au pole antartique – ces .iiii. distances sont toutes equales.

Item: la distance du cecle artique siques au tropique d'esté et celle du tropique d'yver siques au cecle antartique sont ensembles equales. Mais chascune de ces .ii. distances est plus grande que nulle des autres quatre pres de la motié. Car la distance de l'equinocial au tropique d'esté est .xxiii. degrés et demi ou environ, et celle est la distance du cecle artique jusques au pole. Ce sont ensemble aussi comme .xlvii. degrés, lequels sont a oster du quartier du cecle d'entre l'equinocial et le pole, c'est assavoir de .xc. degrés, si demeure environ .xliii. degrés.

Ce est la distances d'entre le tropique d'esté et le cecle artique. Et tele distance a il entre le tropique d'yver et le cecle antartique, si comme dit est.

460 .ii.] 2 DP; deux BHMN 461 Et est l'un] Donc cellui qui circuyt le pole artic est M l'un] *om.* D 461-462 qui ... artique] *om.* M 462 nous] *om.* D 463 est ... vers] circuyt M sous] *add.* nous et B vers] entour D antartique] artique H; antratique EC; entartique G 463-464 est ... antartique] *om.* B 464 antartique] antratique EC; entartique G .iiii.] 4 DP; quatre M; deux B pareillies ou] *illeg.* D; appellés paralelles ou M 465 .i.] un/ung ABDFGHKMNP ou en partie] *om.* HN en$_{2,3}$] a K 466 et] ce N n'est] *om.* P aussi] ainsi ABDFGHKMN ou] au P 468 jusques$_{1,2}$] siques ABCG celle de] *om.* K 469 antartique$_{1,2}$] antratique EC; entartique G jusques] siques ABCG .iiii.] 4 DP; quatre BMN toutes] *om.* M 470 siques] jusques DHKMNP celle] *om.* HN 471 siques] jusques DHKMNP antartique] antratique EC; entartique G .ii.] 2 DP; deux BHM 472 quatre] .iiii. ABFG; .iii. K pres] presques FG de] que K 473 d'este] *om.* N .xxiii.] 23 DMP demi] ½ M celle] tele/telle GH 474 jusques] siques BAFG .xlvii.] 47 BDMP 476 .xc.] 90 DMP; nonante ABFGHKN .xliii.] 43 BDMP 477-478 d'este ... tropique] *om.* K 478 si] *om.* KN

3 Here and in lines 464, 469 and 471 ms E has "antratique" instead of "antartique". This scribal error is corrected in line 478.

LE .XVIIIE CHAPISTRE. DES ARCS DU JOUR ET DE LA NUIT

Le jour naturel est l'espace de temps en laquelle le soleil fait .i. tour entier entour la terre, si comme quant il vient d'un meridian derechief. Item: a cellui meismez, et le jour artificiel est entre soleil levant et soleil couchant; et la nuit est le demourant.

Donques le soleil on chascun jour naturel descript aussi comme .i. cecle, et n'est pas proprement cecle, mes petit s'en faut. Car le soleil ne revient pas en .i. meisme point du meridian dont il partit, mes .i. petit plus hault ou plus bas, en maniere de vis ou du fil entour la fusee.

Ces cecles ici sont appelés "parallelles", et en y a .iii. tant seulement signes et descrips en l'espere materiel pour ce que il sont les plus principalx. Ce sont les .ii. tropiques et l'equinocial.

De ces cercles dessus dis ou de chascun d'iceuls, la partie qui est dessus l'ozizon est appellee "l'arc du jour" ou "l'arc journel", car tant comme le soleil i est ou y passe, il est jour. Et la partie qui est dessous est appellee "arc nocturne".

Et pour ce, tant a plus grant partie d'un de ces cecles sus l'ozizon, tant est le jour artificiel plus lonc et la nuit plus courte. Et aussi, au contraire, tant mains, tant f. 10^{r} est [le jour] plus / court et la nuit plus longue.

Or appert il asses clerement que l'ozizon droit interseque et trenche tres tous ces cecles par le milieu, par quoy il s'ensuit que touz ceuls qui habitent en tel ozizon – c'est sous l'equinocial – ont en touz temps le jour et la nuit equals.

479 Le ... chapistre] *om.* HK; Le .xviii. c. F Le ... nuit] *om.* G; .xviii.^{e} chapitre N .xviii.^{e}] 18/18me DMP nuit] *add.* .xviii. K; *add.* .xviii. chapitre H 480 en laquelle] ou quel B .i.] un/ung ABDFHKMNP 481 derechief] *interlin.* C Item] *om.* ABDFGHKMN 482 entre] *add.* le BHKNP entre ... couchant] entre le soleil couchant D; le *add. in marg.* soleil levant et D et$_1$] *add.* BDHKN 483 on] en CFGHKMN aussi] ainsi M .i.] un/ung ABDFHKMNP 483-484 et ... cecle] *om.* K; *rep.* B 484 en] a HKMN .i.] un/ung ABDFGHKMNP 485 partit] partist P .i.] un/ung ABDFGHKMNP 486 du] de G; d'un DP entour la] a l'entour d' une M 487 .iii.] 3 DP; trois HKMN 488 descrips] escripz K que il] qu' ilz M .ii.] 2 DP; deux BMN 490 dessus] dessous EP; dessuos B; dessous *corr. to* dessus C 491 l'arc$_1$] arc M jour] *add. and del.* et la partie qui est F jour ... journal] journel ou l'arc du jour K l'arc$_2$] l'ar' E comme] que D le soleil] *om.* G; le so|soleil N 492 ou] *add.* qu'il M 493 tant$_1$] *add.* que G a] *om.* D plus] *om.* P de ces] des P 494 lonc] grant B courte] brefve B au] par le M 495 le jour] *om.* ABCEGHKN 496 Or] *illeg.* M asses] *om.* B trenche] touche K 497 tel] *om.* K 498 sous] assavoir dessoubs M; *add.* en B le ... nuit] les jours et les nuits K

Item: il appert aussi que l'ozizon oblique, comme que il voise, ne puet diviser l'equinocial fors en .ii. moitiés. Et pour ce, quant le soleil est la – ces .ii. fois l'an, une en l'entree de Aries, et l'autre en l'entree de Libra – il est adonques equinoxe et sont les jours et les nuits equals par tout le monde.

Item: tous les autres parallelles dessus dis, c'est assavoir tous generalment excepté l'equinocial, l'ozizon oblique les devise en .ii. parties non equales et tant plus est l'ozizon oblique, tant sont ces parties plus inequales. Et pour ce, en tous temps fors a ces .ii. fois dessus dites, les jours et les nuis sont non equales en ozizon oblique, et tant est plus oblique et il a plus grant difference.

Et pour ce, en Suesce[1] ou en Norvoye les jours sont plus longs en esté et plus cours en yver que il ne sont en France, et plus en France que ils ne sont en Jer[usale]m;[2] et ainsi en diverses regions, selonc ce que l'ozizon est plus ou mains oblique.

499 comme] comment AHKMN il$_2$] *om.* N 500 .ii.$_1$] 2 DP; deux BMN .ii.$_2$] 2 DP; deux BHMN 501 en$_1$] a DKMN en l'entree$_1$] entree H en$_2$] a DHKMN adonques] *om.* B 503 Item tous] *om.* M dessus] des dessus G 504 .ii.] 2 DP; deux BHMN non equales] inequales M 505 tant] *add.* plus HNK ces] les M; ses H parties] *add.* des paralleles M plus$_2$] *om.* HKN inequales] inegales DHKMN 506 .ii.] 2 DP; deux BHM non equales] inequales M ozizon] l'orizon MN 507 et$_2$] est G il a plus$_2$] il y a plus P; plus y a de K; tant plus y a B il ... difference] de tant est la difference plus grande M grant] *om.* HK; de N 508 Suesce] Susce P; Suste G ou en] et M en$_2$] *om.* P 508-509 en$_3$ ... yver] *om.* G 509 il] ilz MNP ils] il FG 510 Jerusalem] Ierl'm CE; Ihrlm BFGK; Ihlam̄ D; Jehrlm̄ AMP ainsi] aussi P selonc] secon/secou C ce] *om.* B est] y est B plus] *om.* N

1 Sweden.

2 Jerusalem's latitude is 31° 47′ (cf. 49° for Paris).

LE .XIX.[E] CHAPISTRE. DU LEVEMENT ET DU RESCONSEMENT DES SIGNES EN L'OZIZON

Tout le zodiaque leve et reconse en .i. jour naturel sus chascun ozizon en terre bien habitable. Et pour ce que il est cecle oblique, il ne lieve pas reguleement en ces parties, aussi comme fait l'equinocial, mes l'une partie lieve plus tost, et l'autre plus a loisir, ou plus belement, en variable maniere. Et ceste variacion est faite .iiii. fois en l'ozizon droit es .iiii. poins principauls du zodiaque. Car la quarte partie du zodiaque, qui est du premier point de Aries siques en la fin de Gemini, et la quarte [partie] de l'equinocial, qui est en droit acosté, commencent a lever en .i. moment, et finent en un moment. Et aussi est il des autres .iii. quars du zodiaque et des autres .iii. quars de l'equinocial.

Mas de cest premier quart au commencement, la partie du zodiaque lieve plus tost et la partie de l'equinocial plus a loisir, et pour ce, est il tousjours en celle partie plus levé du zodiaque que de l'equinocial. Et nient moins, ces .ii. quars, l'un de zodiaque, l'autre de l'equinocial, acomplissent leur levement ensemble.

Et est semblable comme se .ii. hommes aloient de Paris a Saint Denis[1] en .ii. / f. 10v heures, et partissent en .i. moment et arrivassent la en .i. moment, mes que .i. alast plus tost en la premiere heure et passast l'autre, et en la seconde heure se retardast aussi comme en attendant si que l'autre la conseust a terme.

512 Le ... chapistre] *om.* HK; Le .xix. c. F .xix.^{e}] 19/19me DMP du$_2$] *om.* MK 512-513 Le ... l'ozizon] *om.* G; le .xix.^{e} N 513 l'ozizon] *add.* droit DFM; *add.* droit .xix. K; *add* .xix. ca. H 514 leve et] est D .i.] *om.* K; un/ung ABDFGHMNP 515 est] *om.* K 516 aussi] ainsi M fait] *om.* B l'une] une ABFM 517 a loisir] alloisir P .iiii.] 4 DMP; quatre B 518 es] en G .iiii.] 4 DMP 518-519 Car ... zodiaque$_2$] *om.* N 519 du] des le B siques] jusques DHKMNP en] a DKN de$_2$] du F 520 partie] *om.* EP qui ... acoste] a luy correspondant M acosté] encoste F a ... moment$_2$] en ung instant a lever et en ung moment finissent M .i.] un/ung ABDFHKNP 521 aussi] ainsi DMP .iii.] 3 DMP; trois H 521-522 du ... quars] *om.* HN 522 .iii.] 3 P; trois BDM 524 la partie] celle/telle HN l'equinocial] *add.* plus tort ou HKN celle] ceste D; telle M 525 .ii.] 2 P; deux HMN 525-526 que ... zodiaque] *om.* BD 526 acomplissent] *om.* D 527 .ii.$_1$] 2 DP; deux BHMN Saint] Sanct H .ii.$_2$] 2 DP; deux HMN 527-528 en .ii. heures] *om.* B 528 .i.$_1$] un/ung ABDFHKMNP et$_2$ ... moment$_2$] *om.* FKM .i.$_2$] un/ung ABDHNP .i.$_3$] l'un/l'ung ABDFHKMNP 529 plus] *add.* droit DG en$_1$] a K seconde] ii. F; 2^{e} D se] si F retardast] tardast ADFHN 530 si] jusque P; jusques a ce M si que] siques F la conseust] l'eust aconceu M a] au DFGHK

1 Two men going from "Paris" to "Saint Denis" is one of Oresme's favourite images. The distance from Notre Dame de Paris to the Basilica of Saint Denis is a little under 10 km, and modern walking time is estimated at a little over 2 hours.

Il appert clerement que cestui auroit tousjours plus passé la voie que l'autre, et semblablement est il en nostre propos, si comme il appert par le mouvement de l'espere materiel ou artificiel. Et est assavoir que quant aveques .i. signe, qui est la .xiie partie du zodiaque, il lieve plus de la .xiie partie de l'equinocial, l'en dit que il lieve droit. Et quant il y en lieve moins, il lieve obliquement, et aussi de quelcunques partie du zodiaque. Et en la maniere que j'ai dit du premier quart du zodiaque, ainsi est il a entendre du tierz quart, qui est du premier point de Libra siques en la fin de Sagitarius. Mes les .ii. autres quars du zodiaque au contraire lievent au commencement droit. Et tant comment il mettent a lever, tousjours est il plus levé de l'equinocial que du zodiaque.

Item: l'en doit savoir que quiconques .ii. signes opposites en l'ozizon droit ont equales ascensions, c'est a dire que autant comme il lieve de l'equinocial aveques l'un, autant en lieve il aveques l'autre.

Les signes opposites sont Aries et Libra; item: Taurus et Scorpius, item: Gemini et Sagitarius; item: Cancer et Capricornus; item: Leo et Aquarius; item: Virgo et Pisces.

531 appert] *add.* tout B plus] *om.* B la] de la ADFGHKMN 533 ou artificial] *om.* B .i.] un/ung ABDFGHKMN; ung ung P qui est] il lyeve plus de M 533-534 qui ... zodiaque] *om.* M; *marg.* A 534 .xii.$^{e}_{1}$] 12 D; 12me P zodiaque] *add.* que B il] *om.* HN .xii.$^{e}_{2}$] 12^{e}/12me DP 534-535 que il] tel signe FN; que tel signe BGHK; que tel signe est la 12^{e} partie du zodiac M 535 y] *om.* DGKNP; luy BM en lieve] est leve D il$_2$ ... obliquement] *om.* B aussi] ainsi M 536 que] comme D du premier quart] premierement [*lacuna/erasure*] *corr. to* premierement du quart D 537 du$_1$] *om.* K tierz] 3 D quart] M *inserts ll. 585-629 (Chap. 21):* "Mais en l'ozizon ... ne ne reconssent." qui] et M point] degre D 538 siques] jusques DHKMNP en] a KMN les] *om.* B .ii.] 2 DP; deux BHNM 539 comment] comme FK; que N il$_1$] ilz PHMN 540 plus] *om.* N 541 quiconques] quelconques AGK .ii.] 2 DP; deux BHMN 544-546 Aries ... Pisces] Aries et Libra; Taurus et Scorpio; Gemini, Sagitarius; Cancer, Capricornus; Leo, Aquarius; Virgo et Pisces DM

LE .XX.[E] CHAPISTRE. DU LEVEMENT DES SIGNES EN L'OZIZON, OU EN L'ESPERE OBLIQUE

Aussi comme l'ozizon oblique interseque [et] divise l'equinocial en .ii. moitiés aussi divise il le zodiaque en .ii. moitiés, si que sus l'ozizon oblique a tousjours la moitié du zodiaque.

Et pour ce, en chascun jour artificiel, tant soit brief ou long, il lieve sus l'ozizon oblique la moitié du zodiaque – ce sont .vi. signes ou l'equipollent, ne plus ne mains precisement et autant de nuit.

Pour quoy il couvient que es jours qui sont plus briefs que les nuis, les signes lievent plus tost et es autres jours plus a loisir, et ainsi le zodiaque ne lieve pas regulieement en ses parties comme fait l'equinocial, mais moins reguleement assés que en l'ozizon. Et ainsi comme ils avoiet .iiii. fois variacion en l'ozizon droit, il a .ii. fois, tant seulemen[t], variacion en l'ozizon oblique; car la moitié du zodiaque, qui est du commencement d'Aries siques en la fin de Virgo tout ensemble met autant
f. 11r de temps a lever comme la moitié de l'e/quinocial qui est decosté soi. Et commencent a lever en .i. moment et finent en .i. moment. Mais ceste moitié du zodiaque lieve au commencement plus tost et ceste moitié de l'equinocial plus a loisir assés, et c'est appellé "lever obliquement". Et tant comme il mettent a lever, tousjours est il plus levé du zodiaque.

Item: l'autre motié du zodiaque, qui est du commencement de Libra siques en la fin de Pisces, et la moitié de l'equinocial, qui est encosté soi, commencent et lessent a lever ensemble. Mais l'equinocial en celle partie lieve au comencement

547 Le ... chapistre] *om.* H; Le .xx. c. F .xx.[e]] 20/20me DMP levement] reconssement H 547-548 Le ... oblique] *om.* G; Le .xx.[e] c. N; Du levement et recousement des signes de l'ozizon ou espere oblique DL 548 ou ... oblique] .xx. chap. H; *add.* .xx. L en] est P en l'espere] espere M 549 Aussi] ainsi M et] *om.* EP l'equinocial] le zodiaque B .ii.] 2 DP; deux BHM 550 aussi ... moities] *om.* BH .ii.] 2 DP; deux BHMN si que] siques ACF 552 il] *erased* D 553 moitie] ½ M .vi.] 6 DP; six HM ou] en N 554 precisement] *om.* B 555 Pour quoy] Parquoy M que les nuis] *om.* B; *add.* que K 556 jours] *om.* BDFKM ainsi] ausi F 557 en ... reguleement] *om.* K asses] ainsi ADFHKN 558 l'ozizon$_1$] *add.* droit BCDHKN ainsi] aussi M .iiii.] 4 DP; quatre BHM 559 .ii.] 2 DP; deux BHMN tant seulement] *om.* B; tant seulemen E 560 siques] jusques DHKMNP 562 .i.$_1$] un/ung ABDFHKMNP moment$_{1,2}$] mouvement P et ... moment$_2$] *om.* D .i.$_2$] un/ung ABFHKMNP 563 et$_1$] en M de l'equinocial] *om.* P 564 il$_1$] ilz PHMN 566 Item] *corr. from* et C siques] jusques HKMNP en] a M 567 est encoste] encostoste K; est a l'endroit de M

plus tost, et le zodiaque plus a loisir. Et c'est appellé "lever droit" et est tousjours en celle partie plus levé de l'equinocial que du zodiaque. Et neantmoins ces .ii. moitiés finent leur levement ensemble. Et pourroit estre declairé par l'exemple qui fus mis ou chapistre precedent de .ii. hommes qui vont a Saint Denis.[1]

Item: la moitié du zodiaque du commencement de Cancer sique en la fin de Sagitarius, en levant, aporte plus de la moitié de l'equinocial avec soy. Car en tant comme elle met a lever, il lieve plus de la moitié de l'equinocial, si que ceste moitié toute ensemble lieve droit, et l'autre moitié du zodiaque lieve obliquement.

Tout ce qui est dit en ces .ii. chapistres appert visiblement par example en l'espere artificiel.

569 tost] *om.* C a loisir] alloisir P est] en F 570 .ii.] 2 DP; deux BHMN 571 finent ... levement] lyevent M leur lievement] leurs levemens K Et ... declaire] Et ce peult estre entendu M 572 fus] est M ou] au P .ii.] 2 D; deux BKMNP Saint] Sanct H Denis] Denys A; *add.* qui est a deux lyeues de Paris M 573 siques] jusques DFHKMNP en] a MN 574 plus] *om.* F 574-575 avec ... l'equinocial] *om.* K 575 met] *om.* H si que] siques GM; *add.* toute G 577 .ii.] 2 DP; deux BHMN par example] *om.* M

1 Chap. 19, ll. 527-530.

LE .XXIE CHAPISTRE. DES .II. AUTRES MANIERES DU LEVEMENT DES SIGNES

Les [signes] du zodiaque lievent [en] aucuns liex selonc leur ordre naturel, l'un aprés l'autre, ce est assavoir Aries devant Taurus, et Taurus avant que Gemini, et aprés ensuiavment. Et ainsi lievent il et reconsent en l'ozizon droit et en l'ozizon oblique sous le ciel entre l'equinocial et le cecle artique. Et de ce est a entendre le chapistre precedent. Mais en l'ozizon de ceuls qui habitent sous le cecle artique, il avient chascun jour naturel une fois par le mouvement journel du zodiaque et de ses poles que le pole du zodiaque est aveques leur cenith ou que le pole du zodiaque est leur cenith par .i. seul instant. Et est instant .i. moment de temps indivisible qui est tantost passé et soudainement. Et en cil moment leur ozizon et l'ecliptīcle est tout .i. ou ensemble. Et ainsi le zodiaque ou l'ecliptique est leur ozizon. Et en cil moment nul des signes du zodiaque ne est sus leur ozizon ou levé, ne il [n'en a] nul dessous
f. 11^{v} leur ozizon ou reconsé, mais tantost aprés cil moment sans / moien il ha .vi. signes sus leur ozizon ou l'equipolent et .vi. dessouz.

Et ainsi a ces ici qui habitent sous le cecle artique, comme dit est, la moitié du zodiaque leur lieve lors soudainement et tout ensemble. Et aprés cil moment il reconse partie de la moitié levee, et partie de l'autre moitié lieve ordeneement, l'un

579 Le ... chapistre] *om.* HKN; Le 21/21me chappitre DMP Des] De ABN .ii.] 2 DP; deux BHMN autres] *om.* ADHKM 579-580 Le ... signes] *om.* G; *add.* .xxi. chap. H; *add.* .xxi. K; *add.* .xxie c. N 581 signes] figures ECP zodiaque] *add.* qui D en] *om.* E selonc] secon/secou C 582 l'autre] *om.* M ce est assavoir] *om.* D devant] avant que K que] *om.* BN 582-583 apres ensuiavment] ainsi des aultres ensuyvans M 583 ensuiavment] ensuivant FP ainsi] aussi DF reconsent] escousent M en$_2$] *om.* D oblique] *add.* as h[ab]itans ABF; *add.* aux habitans DGM; *add.* aux habitacions HN; *add.* es habitacions K 584 est] en D a] *om.* B 584-585 le ... precedent] *om.* M 585 avient] advient K 585-629 Mais ... recoussent] *moved to line 537 (q.v.)* M 586 une fois] *om.* F ses] ces H 587 est aveques] passe par M; et avec B aveques ... est$_2$] *om.* P est$_2$] *add.* en D 587-588 ou ... cenith] *om.* BFHM 588 .i.$_1$] un/ung ABDFGHKNP est$_1$] *om.* K; cest M; *add.* un B est instant] en D .i.$_2$] un/ung ABDFHKNP .i. moment] ou mouvement M; un mouvement KN temps] *add.* est M qui est] et M 589 tantot] toust B .i.] un/ung ABFGHKNP 589-590 est tout .i. ou] sont M 590 ou$_1$] *om.* K ainsi] aussi F Et$_1$ ... ozizon] *om.* M le zodiaque ou] *om.* G ou$_2$] et K est] en B; et P en] *om.* P 590-592 cil ... reconse] cellui temps nul des signes n'est dessus ou dessoubz l'orizon M 591 leur ozizon] l'orizon P ou leve] esleve N n'en a] neua EC; ne na P; n'a BDG; ne a AF 592 leur ozizon] *om.* K ou] ne N cil] icellui P cil ... ha] et subitement ils sont M .vi.] 6 DP; sex M 593 leur ozizon] l'orizon N .vi.] 6 DP; sex signes M 594 ces ici] ceux ici N; ceulx yci H; ceulx cy MP habitent] h[ab]itent F 594-595 Et ... lors] Et l'autre moite leur reconsse H 595 ensemble] *add.* et l'autre moitie leur reconse soudainement et toute ensemble ABFGHMN cil] *om.* G; icellui P cil moment] *om.* D 596 et ... lieve] *om.* GM l'autre] la N

signe aprés l'autre, aussi comme il est dit ou chapistre precedent de autre ozizon oblique.

Item: a ceulz qui habitent entre le cercle artique et le pole artique, une partie du zodiaque est tousjours sous leur ozizon – de laquelle le premier point de Capricornus est le milieu – et celle partie ne lieve onques sus leur ozizon. Semblablement, une autre partie du zodiaque, equale a la partie dessus dite, est tousjours sus leur ozizon sans reconser; et de celle partie le premier point de Cancer est le milieu. Et ainsi demourent .ii. autres parties du zodiaque dont l'une est devers Aries, et l'autre devers Libra, lesqueles lievent et reconsent en tel ozizon. Et est assavoir que la partie devers Aries, et de laquelle le premier point d'Aries est le milieu en tel ozizon, lieve a rebours, et ce devant derriere, non pas selonc l'ordre naturel des signes. Car illeuques lieve Taurus avant que Aries, et Aries avant que Pisces, et toutes voies ceste partie et ces signes meismes reconsent illeuques selonc leur ordre naturel. Car Pisces reconse avant que Aries, et Aries avant que Taurus. Et de l'autre partie du zodiaque qui est a l'opposite, de laquelle le premier point de Libra est le milieu, il est tout a la converse et a maniere contraire. Car les signes en tel ozizon lievent selonc leur ordre naturel, si comme Virgo avant que Libra, et Libra avant que Scorpio, mais il li reconsent a rebours, ce devant derriere, si comme Scorpio avant que Libra, et Libra avant que Virgo.

Item: a ceuls qui habitent droit sous le pole, leur ozizon est l'equinocial et ont en tous temps les .vi. signes, septentriennes sus leur ozizon, c'est assavoir Aries, Taurus, Gemini, Cancer, Leo, Virgo. Et les autres .vi. signes ou l'autre moitié du zodiaque est tous temps sous l'ozizon. Et ainsi les signes n'i lievent ne ne reconsent.

597 l'autre] la N aussi] ainsi DM dit] *om.* N precedent] *om.* ADG precedent ... ozizon] cy-devant en parlant de l'orizon M 599 a] *om.* B et ... artique] *om.* N 600 leur] le D 601 et ... onques] de celle partie et icelle jamais ne monte M onques] point N leur ozizon] l'orizon N 601-603 une ... milieu] l'autre partie du zodiac dont le pemier point de Cancer est le milleu d'icelle jamais ne se reconse soubs leur orizon, mais est tousjours dessus icelle. M 603-604 ainsi ... .ii.] autre *illeg.* et D 604 demourent] demeurent AGHKMN .ii.] deux HBMNP 605 tel] leur M est] *om.* N 606 partie] *add.* de P et ... d'Aries] *om.* K 607 a rebours] arrebours P et] c'est M selonc] selon C l'ordre] leur ordre D des] telz F 608 illeuques] illes M avant que$_1$] devant D 609 illeuques] illeqs | illeques F selonc] secon/secou C 611 premier point] promer G le$_2$] ou K 612 il] et K a$_2$] *om.* M; en HK; *add.* la F signes] *add.* d'icelle partie M tel] cel G; cil HN 613 selonc] secon/secou C leur ordre] l'ordre K Libra$_1$ ... que] *om.* G 614 Scorpio] Scorpius AF *and elsewhere*; Scorpie G il li/ly] il DK; il y BG; ilz PHN; ilz y M si] *om.* HK 615 et ... Virgo] *om.* B 616 droit] *om.* M est l'equinocial] et l'equinoctial sont tout ung M et] en B; si D 617 temps] *om.* AG .vi.] 6 DP; sex M septentriennes] septentriones AF; egaulx D c'est assavoir] .s. [=scilicet?] D 618 .vi.] 6 DP; sex M ou l'autre] en une D 619 zodiaque] *add.* que D tous temps] tousjours BM l'ozizon] leur orizon DFGH ainsi] aussi DP n'i lievent] d'icelle moictie ne lyevent M ne ne] ne K reconssent] *add.* point M

Item: selonc ce qu'il est dit de ceste partie du monde desa l'equinocial, f. 12r semblablement / puet on dire de l'autre partie qui est par dela l'equinocial devers le pole antartique, quant a tous les poins touchies en cest chapistre, fors tant que les uns signes apparoissent par desa, et les autres par dela; les uns lievent ordeneement par desa, et a rebours par dela, si comme on puet veoir clerement qui entent les choses dessus dites.

Je di donques, en recapitulant par ordre retrograde, que en aucuns lieus les signes du zodiaque ne lievent ne ne reconsent; et en aucuns lieus, les uns lievent a rebours et reconsent par ordre; les autres lievent par ordre et reconsent a rebours; les autres ne lievent ne ne reconsent. Et en aucuns lieus il lievent les uns en aucunes fois soudainement; les autres reconsent soudainement. Et en aucuns lieus il lievent tous, et tous reconsent ordeneement, l'un aprés l'autre, et partout la ou il lievent et rescousent, les uns lievent droit, et les autres obliquement. Et tout ce que est dit en ces .iii. chapistres du levement et reconsement des signes appert visiblement par example en l'espere artificiel, comme dit est.

620 selonc ce] tout ainsi M; secun ce C qu'il est] que D de] en K monde] moment K; *add.* qui est M desa] de K; de cha^{la} D 620-627 Item ... reconsent] *om.* N 621 peut on] peult l'en M par] *om.* M 622 touchies] touchans H tant] *add.* seulement 623 apparoissent] apperent K 624 et] *add.* lievent K et a rebours] *erasure and add.* a tenets(?) D si ... entent] *om.* K on] l'en B 624-625 puet ... dictes] peult veoir clerement en l'espere materielle. M 625 dessus] *om.* B 627 signes ... les] *om.* N 629 les$_1$ ... reconsent] *om.* B ne] ny P Et] Car M en$_1$] *om.* G aucuns] autres K il] *om.* N uns en] *om.* B en$_2$] *om.* ACM 629-630 il ... fois] les uns aucunefois lievent H; les uns encore lievent K 629-631 les ... tous$_2$] et D 630 les ... soudainement$_2$] *marg.* M; *om.* K aucuns] autres K il] ils/ilz PBHMN 631 tous$_1$] *om.* M il] *interlin.* C et$_3$] ilz N 631-632 ordeneement ... rescousent] *om.* K 633 .iii.] 3 DP; trois/troys BHM; deux N 633-634 par example] *om.* M 634 comme dit est] *om.* D; *add.* etc. H

LE .XXIIE CHAPISTRE. DE L'EXCENTRIQUE DU SOLEIL

Tout cecle qui divise son espere en .ii. moitiés et a son centre ou centre du monde est dit "conce[n]trique". Et tout cecle qui divise l'espere en .ii. moitiés et ne a pas son centre ou centre du monde, est appellé "excentrique".

Or est il ainsi que le cecle que le soleil descript en son espere sous le zodiaque n'est pas concentrique, mais a son centre aillours que ou centre du monde, et est
f. 12^{v} excentrique, si comme il appert par exemple en ceste figure.[1] Et le point / de cest cecle qui est plus haut et plus loings du centre du monde est appelé "*aux solis*".[2] Et l'autre point contraire qui est plus bas et plus pres du centre du monde, est appellé "*oppositum augis*".[3] Et est asavoir que *aux solis* est une fois en une endroit du zodiaque, et autre fois en autre, selonc le propre mouvement de le .viiie espere qui est moult tardif. Et dient les astrologiens que il passe en .c. ans un degré du zodiaque. Et est maintenant *aux solis* vers la fin de Gemini, pres du premier point du Cancer.[4] Pour entendre la maniere comme cest excentrique puet estre, les philosophes dient que toute l'espere du soleil, de quoy il est mencion ou quart chapistre,[5] est devisé en .iii. parties, desquelles l'une est moienne et excentrique et

635 Le ... chapistre] *om.* HK Le ... soleil] *om.* G; *add.* .xxii. K; *add.* .xxii. chap. H .xxiie] 22/22me DMP chapistre] *om.* A 636 qui] *add. and del.* est E .ii.] 2 DP; deux BMN moities] *om.* G centre ou] *om.* A 637] concentrique] concetrique E .ii.] 2 DP; deux BHM moities] menieres G pas] *om.* N 638 ou centre du] au P 640 mais] *om.* HN; car il K 641 figure] *add.* cy dessus faicte HN; *add.* prochain ensuyvant M 642 cecle] *add.* precedent B; figure F est$_1$] *add.* le M plus$_1$] au HN plus$_1$ ... monde] en bas K et$_1$] est N plus$_2$] le plus D est appele] *om.* A Et$_2$] Or D 644 une$_1$] aucune K une$_2$] autre K 644-645 en ... fois] *om.* D 645 autre$_1$] aucun K en] *add.* ung P selonc] secun C le .viii.^{e}] la 8me DP; la huit^{em} M 646 .c.] 100 P; cent BFM 648 comme] comment GN cest] est K 649 il est] est fait M; il est fait DK quart] .iiii. G 650 .iii.] 3 D; trois/troys BFNP

1 See Diagram: Chapter 22(A).

2 *Aux solis*: now known as the sun's apogee (or the earth's aphelion). *Aux, augis* is from the Arabic *auj* (اوج), "apogee"; not found in classical Latin.

3 *Oppositum augis*: now known as the sun's perigee (or the earth's perihelion).

4 It is difficult to correlate this position of the sun at apogee with a precise date, given the length of the cycle of rotation of the line of apsides joining the sun's apogee and perigee through the zodiac (between 23,000 and 26,000 years). Ancient astronomers used 15° Gemini as the position of apogee; today it is around 14° Cancer. "The end of Gemini," or "the first point of Cancer" would be reasonable for 700 years ago.

5 Chap. 4, ll. 153-157. Given that each sphere (including that of the sun) has the same thickness throughout, Oresme is probably suggesting that the sphere of the sun is divided into three layers, two of which are shaped such that they cause the middle one to be eccentric.

d'une meisme espesseur ou profundeur en chascune de ses parties; et en icelle est le soleil fichíé et se meut aveques elle.

Dessous ceste ici a une autre partie qui est espesse endroit *aux solis*, et tenve de l'autre part. Et sa superfice concave est concentrique, mais sa superfice convexe est excentrique.

Item: dessus la moienne espere, dont je ay dit devant, il a aussi une espere bossue laquelle est espesse vers *oppositum augis*, et est tenue vers *aux solis*, et sa superfice concave est excentrique, et sa superfice convexe est concentrique, si comme il appert par example en figure.[6]

Et ces .ii. esperes qui sont l'une dessous la moienne, et l'autre dessus, sont aussi comme pour raemplir le lieu, car c'est impossible selonc nature qu'il soit riens vieu.

Toutesvoies, le mouvement du soleil fust aussi bien gardé en metant que il eust epicicle. Mes telles difficultés et des excentriques et des epicicles des autres planetes appartiennent a la theorique des planetes. Et je n'en vueil plus ici parler, mais je ay dit seulement de l'excentrique du soleil, car je en auray a faire aprés.

651 chascune de] toute B ses] ces GHKM 652 se] *om.* D elle] *add.* secunda figura B 653 ici a] cy y'a H; cy en a N endroit] a l'endroit de M 654 concentrique] excentrique F sa] la GM 656 Item] Et M a] ya P aussi] *om.* ADFG 656-657 dont ... laquelle] en a une aultre qui M 657 bossue] *om.* B est tenue] tenue MN; tenvie H sa] la G 658 superfice$_1$] perficie N concave ... superfice$_2$] *om.* M sa] *om.* B; la sa G concentrique] excentrique F 659 par] *add.* ceste D par example] *om.* M en] *add.* ceste K; *add.* la B; *add.* sa M; ou D 660 ces] en ces H; en ses N .ii.] 2 P; deux BHM; *add.* poins K sont] *om.* HN 660-665 Et$_1$ ... apres] *om.* D 661 selonc] par F; secon/secou C riens] *om.* B; nul N vieu] lieu vuit N; vuyt; et toutes ces troys esperes prises ensemble sont une espere de une mesme espesseur en tous endroi, et si est concentrique M 662 soleil] *marg.* P fust] fu H; futN; feust K garde] regarde K que il] *add.* y N 662-664 fust ... planetes] eust peu estre contemplé par epycicle qui eust voulu assi bien que l'on fait par la maniere dessus dis. Mais telles consideracions et difficultés M 663 epicicle] eclipse HN Mes] *om.* G epicicles] eclipses HN 663-664 et$_2$ ... planetes] *om.* B 664 Et ... parler] *om.* B; parquoy de ce nén veul plus parler M n'en] ne B 664-665 Et ... apres] *om.* HKN 665 apres] *om.* M

6 See Diagram: Chapter 22(B).

DIAGRAM: CHAPTER 22(A)[7]

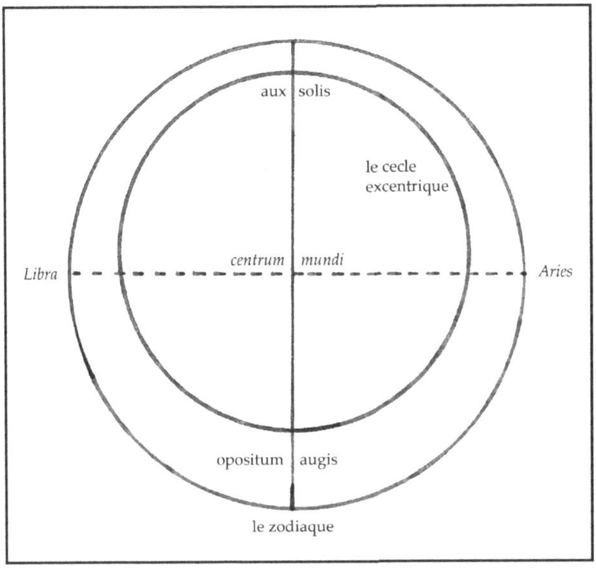

Complete diagram: E CP BFHKMN
Outline / Partial diagram: G
No diagram (or space only): D

aux solis] aux BF   le cecle excentrique] *om.* BFHKN   Libra] BFHKN   centrum mundi] BFHKN   Aries] BFHKN   opositum augis] *om.* H   le zodiaque] *om.* BFHKMN

7 Reference: line 641. Mss ECP form an identical set, with M close to it; mss BFHKN form a separate parallel set.

DIAGRAM: CHAPTER 22(B)[8]

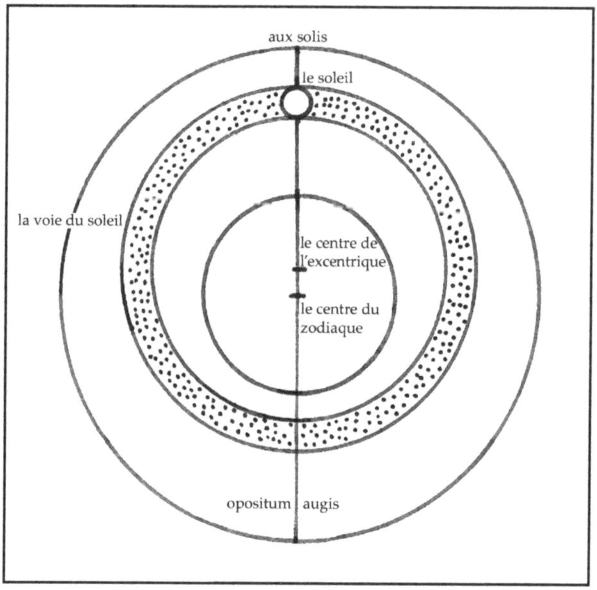

Complete diagram:[9] E CP BFHKN
Outline / Partial diagram: G
No diagram (or space only): DM

aux solis] *om.* FHKN; aux B   le soleil] *om.* BHN   la voie du soleil] *om.* K; l'espere excentrique F; l'espere excentrique du soleil BHN   le centre de l'excentrique] *om.* BFHKN   le centre du zodiaque] *om.* BHKN; le centre du monde F   opositum augis] *om.* FHKN

8 Reference: line 659.

9 Mss ECP form an identical set; mss BFHKN form a separate parallel set.

f. 13r /

LE .XXIII.[E] CHAPISTRE. DE LA PLUS GRANT MO[I]TIÉ DE L'AN

A parler proprement, c'est impossibile que la moitié d'une chose soit plu[s] grande que l'autre; mais, a parler improprement, les astrologiens appelent une moitié de l'an le temps que le soleil met a passer la moitié du zodiaque, du commencement de Aries par Cancer siques en la fin de Virgo. Et en l'autre moitié de l'an, il passe l'autre moitié du zodiaque.

Or est ainsi que le soleil en son excentrique est meu regulierement, et en passe chascun jour .i. degré et petit [moins].[1]

Et qui trairent une line ou .i. dymettre du premier point de Aries siques au premier point de Libre, parmi le centre du monde, il appert clerement que plus de la moitié de l'excentrique du soleil est vers Cancer.

Donques couvient il que le soleil passe plus de la moitié de son excentrique ou f. 13v temps que il va du premier point de Aries / siques au premier point de Libra ou au derrenier de Virgo. Et pour ce couvient il que il y mette plus que la droite moitié de l'an. Et cil temps est appelé la plus grant moitié de l'an et est plus grande que l'autre partie de l'an environ .viii. jours natureus ou plus.

Et qui voudroit appeler toute ceste partie "esté," et toute l'autre "yver", si comme font aucuns aucteurs, il s'ensuirent que esté seroit de tant plus long que yver.

666 Le ... chapistre] *om.* HKN   Le ... l'an] *om.* G; *add.* .xxiii. K; *add.* .xxiii. c./chap. HN   .xxiii.[e]] 23me DMP   moitie] motie E   667 c'est] *add.* chose FM   soit] *add.* plus moins ou B   plus] plu E   plus grande] greigneur HKN   669 le temps] *om.* F   670 siques] jusques DHKMP   en$_1$] a MN   en$_2$] *om.* B   672 est$_1$] *add.* il BKM   en son] *marg.* F   672-673 regulierement ... moins] *om.* M   passe ... jour] chascun jour en passe K   673 .i.] un/ung PABDFGHKN   moins] plus CE; plus *corr. to* moins A   674 trairent] tyreroit M   ou .i. dymettre] *om.* M   .i.] un/ung ABDFGHKNP   siques] jusques DHKMNP   675 point] *om.* B   appert] apparroit DK   676 de ... soleil] *om.* B   vers] et seroit devers K   677 Donques] *rep.* M   excentrique] *add.* que F   ou] en N   678 que il] qui H   va] *om.* K   Aries] *add* par Cancer M   siques] jusques DHKMNP   679 Virgo] *corr. from* Libra B   y] *om.* BDFM; *interlin.* A   que$_2$] de BDFGKM   droite] *om.* D   680 est$_2$] *add.* la B   681 partie] moitie N; moitie ou partieB   de l'an] *om.* D   environ] *add.* de B   .viii.] 8 DMP   jours] heures HKN   natureus] *om.* FM   682 partie] moitie K   toute$_2$] *om.* K   l'autre] *add.* partie D   682-683 si ... aucteurs] *om.* B   683 yver] *add.* cy ensueuse(?) la figure B

1 See the note added to the addenda to this chapter.

DIAGRAM: CHAPTER 23

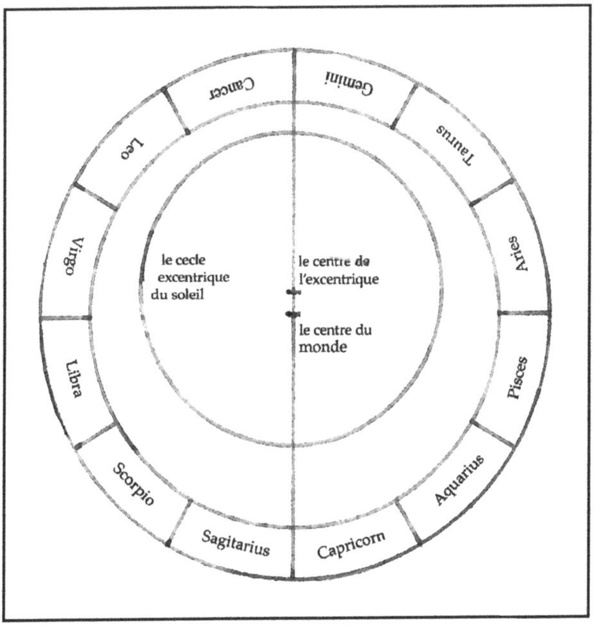

Le cecle du zodiac

Complete diagram:[2] E CP BFHKN
Partial diagram: G
No diagram (or space only): DM

Taurus] *om.* CFHKN Gemini] *om.* CFHKN Cancer] *om.* C Leo] *om.* CFHKN Virgo] *om.* CFHKN Libra] *om.* B Scorpio] *om.* BCFHKN Sagittarius] *om.* BCFHKN Capricornus] *om.* BCFHKN Aquarius] *om.* BCFHKN Pisces] *om.* BCFHKN le cecle ... soleil] *om.* BFHKN; le cecle excentrique du monde P le centre de l'excentrique] *om.* BFHKN le centre du monde] *om.* BFHKN Le cecle du zodiac] *om.* BFHKN

2 Mss ECP form an identical set; mss BFHKN form a separate simpler set.

Addenda: Chapter 23

The different lengths of the seasons are a result of the elliptical orbit of the Earth around the Sun, although in Oresme's time, this is reversed as the "Sun's orbit of the Earth". As in the figure, when the Earth is at the March equinox, we "view" the Sun against the Pisces/Aries constellations behind it. Similarly, when the Earth is at the September equinox, we view the Sun against the Virgo/Libra constellations behind it.

When the Earth is at the June solstice, we view the Sun against the Gemini/ Cancer constellations; and when the Earth is at the December solstice, we view the Sun against the Sagittarius/Capricorn constellations.

Following Kepler's Second Law of Planetary motion (the area swept by a line segment between the moving planet – e.g. Earth – and the fixed Sun is constant over identical units of time), the Earth covers less of its orbit during each day of its longer equinox-equinox path when it is further from the sun, that is, during the March-June-September orbit, and therefore needs more days to travel from equinox (March) to equinox (September). During the September-December-March period, it travels faster and therefore needs fewer days to travel from the September equinox to the March equinox.

Modern calculations set the April-September period as 186 days 10 hours and the September-April period as 178 days 20 hours (a total of 365 days 6 hours). Hence the former is 7 days 14 hours longer than the latter which is Oresme's "about 8 days" (line 681).

(Of course we do not see the constellation in which the Sun happens to be at midday of any one day, because of the blinding light of the Sun itself. But we can see which constellation is in the middle of the sky at midnight, and then calculate the opposite constellation of midday which defines the position of the Sun in the zodiac for that day. For instance, when the we see the constellation of Libra at the middle of the midnight sky, we know that "the Sun is in Aries".)

NOTE: Lines 672-673 are a bit unclear. Disregarding the elliptical nature of the orbits, if the sun travels through 360° in 365 days, it would travel through less than a degree per day (the average throughout the year would be 0° 59' 11" per day). However, between September and April, when it passes through 180° of the equatorial circle in 178 days 20 hours, it would pass on average through very slightly more than 1 degree in one day (about 1° 0' 23").

LE .XXIIII. CHAPISTRE. DE L'INEQUALITÉ DES JOURS NATUREX

Le jour naturel est de soleil levant siques a lendemain soleil levant, ou de midi siques a midi. Et en celle espace de temps le ciel fet un tour, et encores aveques cil tour, autant comme le soleil a passé du ciel par son propre mouvement au contraire du mouvement journel qui est mesuré par l'equinocial, si comme fu dit au secont chapitre.[1]

Or est il ainsi que le soleil passe du ciel par son propre mouvement en .i. jour naturel plus, et en autre, moins, pour .ii. causes principaus: l'une est pour ce que aucuns signes lievent droit et les autres obliquement, tant pour l'obliquité ou pour f. 14^{r} le biais du zodiaque / que pour l'obliquité de l'ozizon; l'autre cause est pour la raison de l'excentrique du soleil.

Et afin que la chose appere plus clerement, poson que le soleil par son propre mouvement passast en chascun jour naturel .i. signe. Je di que le jour ou il passeroit Aries seroit plus brief que celui ou il passeroit Libra en ozizon oblique, comme est le nostre.

Car en cil jour il couvendroit que tout l'equinocial levast sus l'ozizon et encor autelle partie derechief comme il en lieve aveques Aries. Et pour ce Aries lieve obliquement, il ne lieve pas aveques lui la .xii.^{e} partie de l'equinocial mais moins. Et ou jour que le soleil passeroit Libra, il couvendroit que tout l'equinocial levast et derechief encore plus que la .xii.^{e} partie de lui pour ce que Libra lieve droit.

684 Le ... chapistre] *om.* HK Le ... naturex] *om.* G .xxiiii.] 24/24me DMP naturex] *om.* M; *add.* xxiiii. K; *add.* .xxiiii. c./chap. HN .xxiiii. 24/24me DMP 685 de$_1$] des le G siques$_1$] jusques BDKMP siques$_2$] jusques DKMP 686 a midi] *om.* HN le ciel] *marg.* C un] .i. G tour$_1$] *add.* d'orient par occident en retornant vers orient qui est dit mouvement journal M cil] icelle P; ycil H 687 autant] tant G par] de D mouvement] *add.* qui est M 688 mouvement] *add.* de D; *add.* du jour ał N journel ... l'equinocial] dessus dis car il est de occident vers orient M comme: *add.* il ABDFGHKMN secont] ii. F; 2 D 690 .i.] *om.* HN; un/ung ABDFKMP 691 et en] une fois que N en] *add.* ung DM .ii.] deux BHMNP; 2 D est] *om.* K; il est D 692 pour$_1$] *om.* M l'obliquite] obliquité E pour$_2$] *om.* HMN 692-693 ou pour le] et K; ou B 693 que] aussi M; et K pour$_1$] *om.* F l'obliquite] obliquite EP pour$_2$] par N la] *om.* K 694 de] *add.* l'obliquité du soleil de M 695 chose] soleil N plus] *om.* B le] *om.* H par] en M 696 en] *om.* M jour$_1$] *om.* MK .i.] un/ung ADFGHKMNP 697 en] ou F ozizon] l'orizon M 697-698 comme ... nostre] *om.* M 699 cil] icell~ P 700 autelle] autant ABDGHN partie] *om.* ABDFGHKMN comme] *om.* K ce] *add.* que DFGHKM 701 .xii.^{e}] 12/12me DMP 702 levast] *om.* B 703 que$_1$] de K .xii.^{e}] 12me DMP

1 All mss refer to Chap. 2, but the motion of the sun around the earth and along the ecliptic is not found there, nor explicitly elsewhere; one might argue that it is found implicitly in Chap. 5 or 7.

Donques seroit le jour naturel plus long notablement quant le soleil passeroit Libra que quant il passeroit Aries.

Et pour ce, maintenant, que le soleil passe chascun jour environ .i. degré du zodiaque et un degré lieve plus tost, et l'autre plus a loisir, et ameine l'un degré aveques soy plus de l'equinocial, et l'autre moins, je di que, ja soit ce que il y ait moins de difference que se le soleil passoit .i. signe tout entier en .i. jour, toutevois y a il difference, et sont selonc ceste cause[2] les jours naturels plus lon[g] vers la fin de septembre que vers la fin de mars.

704 notablement] *om.* K 705 il] le soleil K 706 pour ce] *rep.* N maintenant] *om.* M environs .i. degre] ung degre ou environ M .i.] un/ung ABDFFHNP 707 un] .i. G plus$_1$] *om.* BD l'un] un/ung HKN; autant sont un D 708 que$_1$] om. B ait] a KN 708-709 di ... que] dys que combien que la differance ne soit pas si grande comme M 709 se] *om.* D .i.$_1$] un/ung ABDFGHKMNP entier ... jour] entour K en .i.$_2$] chascun M .i.$_2$] un/ung ABDFHNP 710 y a il] si a yl M sont] *om.* P; *marg.* C sont ... naturels] pour ceste cause sont les jours M selonc] secun C ceste] *om.* BF long] lons E 711 la] *om.* B

2 In ms E it appears that "cause" has been amended to "causes" (but in a different hand?).

LE .XXVE CHAPISTRE. DE L'AUTRE CAUSE DE L'INEQUALITÉ DES JOURS NATUREX

Pour l'autre cause entendre, soit posé que il parte .ii. lignes du centre de l'excentrique du soleil qui voisent siques au zodiaque par devers *aux solis*, et que il ait entre ces .ii. lignes la .xiie partie de l'excentrique.

Item: posé que il ait .ii. autres lignes au contraire qui semblablement compraignent la xiie partie de l'excentrique de l'autre part si comme il appert en figure.

Je di que les .ii. lignes qui sont vers *aux solis* compraignent moins de la .xiie partie du zodiaque, et les .ii. qui sont vers *oppositum aguis* en comprengnent plus de f. 14^{v} la .xiie partie. Et ceci est a supposer comme chose demonstree par / geometrie. Donques s'ensuit par necessité que quant le soleil est vers *aux solis*, et il passe la .xiie partie de son excentrique, il passe moins de la .xiie partie du zodiaque, c'est assavoir moins d'un signe. Et donques quant il passe .i. degré de son excentrique, il passe moins d'un degré du zodiaque. Et de l'autre part vers *oppositum augis*, est tout au contraire, car quant il passe .i. degré de son excentrique, il passe plus d'un degré du zodiaque.

Or est il ainsi que il passe chascun jour environ .i. degré de son excentrique reguleement. Et pour ce, il ne passe pas le zodiaque reguleement.

712 Le ... chapitre] *om.* HK .xxve] 25/25me D De ... l'inequalite] De l'autre inequalite D; D'une autre inequalite K 712-713 Le ... naturex] *om.* G; De l'autre inequalite .xxv. c. N; *add.* .xxv. K; *add.* .xxv. chap. H 714 parte] passe HKN .ii.] 2 DP; deux BGHMN 715 siques] jusques BDHKMNP ait] y ait P 716 .ii.] 2 DP; deux BMH .xii. ^{e}] 12me DMP 717 Item ... il] Soit posé encores qu'il y M pose] posons G il] *add.* y B .ii.] 2 P; deux MHN; .i. D 718 compraignent] comprent M .xiie] 12me MP part] *marg.* C 718-719 si ... figure] *om.* HKN 719 figure] ceste figure D; la figure cy apres ensuyant M 720 di] *add.* doncques HKN .ii.] *om.* D; 2 P; deux BMH .xiie] 12me DMP 721 .ii.] 2 DP; deux BMHN en] *om.* BM 722 .xii.^{e}] 12me DMP partie] *add.* du zodiaque BM 723 s'ensuit] *add.* il AFHGKM que] que que E 724 .xii.$^{e}{}_{1}$] 12^{e} DMP de$_{1}$ ... partie$_{2}$] *om.* M de$_{2}$] *om.* G .xii.^{e2}] 12me DP 724-726 de$_{2}$ ... moins] *om.* N 725 assavoir] *om.* D d'un] du G donques] aussi M il] *add.* est H .i.] un/ung ABDFGHKMP 726 augis] *add.* il M 726-727 moins ... passe] *om.* P 727 il$_{1}$] *add.* est HN .i.] un/ung ABDFGHKMN plus] *corr. from* mains B 729 ainsi] *om.* D jour] *add.* ou H .i.] un/ung ABDFHKMP 730 reguleement$_{1}$] *add.* ou environ K Et ... reguliecment$_{2}$] *om.* D

Et pour ceste cause seroient les jours naturiex plus longs quant le soleil est devers *oppositum augis*, car adonc tout le zodiaque lieve en .i. jour naturel, et plus d'un degré, et moins, quant il est vers *aux solis*.

Et ainsi, selonc ceste cause, seroient les jours naturels plus longs en yver que en esté. Mais aucunefois l'autre cause dite ou chapistre precedent[1] est au contraire.

Toutesvoies, pour toutes ces .ii. causes sont il plus longs vers la fin de septembre que vers la fin de mars. Et nientmoins, la moitié de l'an devers yver est f. 15^{r} la plus petite, comme / il fu dit ou .xxiiiie chapistre,[2] car l'acroissement de ces jours naturels en quantité n'est pas si grant comme l'accroissement de l'autre moitié en nombre de jours.

731 seroient] sont M   732 devers] vers K   adonc] donques K   .i.] *om.* BF; un/ung ADHKMP   733 d'un degre] encores une partie du zodiac M   est] *om.* P   734 selonc] secon/secou C   736 toutes] *om.* K   .ii.] 2 DP; deux BHM   il] ilz FHKMNP   738 il fu dit] dit est B   fu] a esté M   .xxiiiie] 24/24me DP; .xiiii. BMF   739] en quantite] *om.* K

1 Chap. 24, ll. 685-711.
2 This should probably be Chap. 23 (not 24), ll. 667-683.

DIAGRAM: CHAPTER 25

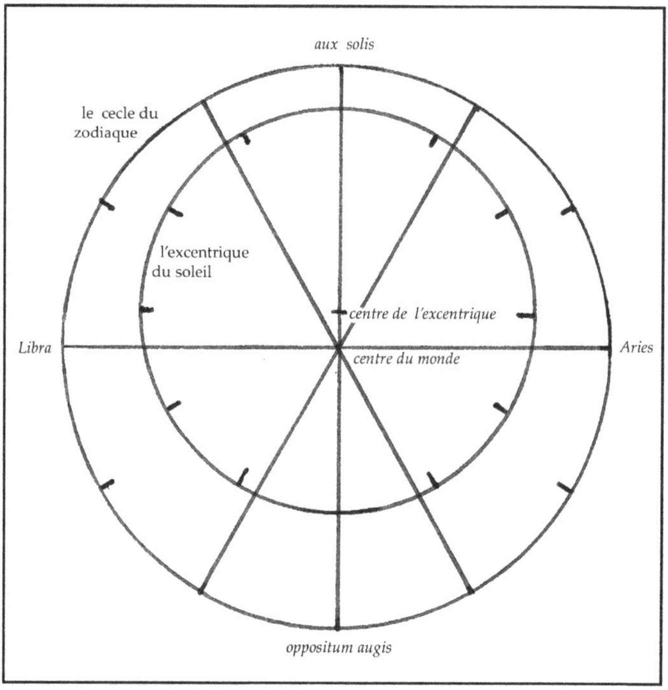

La figure qui preuve l'i[n]equalité des jours

Complete diagram: E BCFHKMP
Partial diagram: GN
No diagram (or space only): D

aux solis] BHM; aux FK l'excentrique du soleil] *om.* BK; excentricus solis M; l'excentrique H; le cecle excentrique du soleil C le centre de l'excentrique] C; centrum excentrici M Libra] C Aries] C le centre du monde] C; centrum mundi M le cecle du zodiac] *om.* BFHKM oppositum augis] HK La figure ... jours] *om.* BCFHKM l'inequalité] l'iequalité E

LE .XXVIE CHAPISTRE. DE LA MESURE DE LA TERRE

Comme il fu dit ou chapistre premier,[1] la terre est aussi comme roonde, et pour ce, selonc ce que va de pais en autre, l'en a autre ozizon et apparoist autre partie du ciel. Et se .i. homme aloit de septentrion droit vers midi, le pole artique li seroit continuelment moins eslevé; et se il aloit au contraire, il li seroit plus eslevé.

Et pour ce, se il aloit droit vers midi sous du meridian, tant que le pole artique fust moins eslevé sus son ozizon par la .xiie partie du meridian, il auroit passee la .xiie partie du circuite de la terre. Et se il aloit tant que le pole li fust moins eslevé d'un degré seulement, ou au contraire, tant que il fust plus eslevé d'un degré, il auroit passé .i. degré du circuite de la terre. Et auroit ou circuite de la terre .ccclx. teles espaces. Et par ceste maniere fut trouvee la mesure de la terre, car l'elevacion du pole ou d'une estoille est aesié a trouver par l'astrolabe ou par le quadran.

Or est il ainsi trouvé par experience que se .i. homme aloit vers septentrion tout droit, tant que le pole li fust plus eslevé d'un degré en droit midi, tant que il fust

741 Le ... chapistre] *om.* HKN Le ... terre] *om.* G .xxvie] 26/26me DMP De ... .xxvi. c.] *rep.* N mesure ... terre] division de la terre en cinq parties B; *corr. from* division de la terre en .v. parties A; *add.* .xxvi. K; *add.* .xxvi. c./chappitre HN 742 fu dit] a esté dit M; fus~~fois~~ P aussi] ainsi M 742-743 la ... de] et pour ce que lén va d'un B 743 selonc] secon/secou C ce] *om.* M que] *add.* l'en ACDFGHMN autre$_1$] l'autre B a] *add.* continuelment K et] qui N apparoist] appert HKM; apperest N 744 .i.] un/ung ABDFHKMNP droit] *om.* KM midi] *add.* droit et sans decliner K seroit] *om.* G 745-746 il$_1$ ... aloit] *om.* D 746 droit] *om.* N du] en B; .i. DG; un/ung FHKMN tant] en tant B 747 moins esleve] plus esleve mais B son ozizon] l'orizon M .xii.^{e}] 12/12me DMP; *corr. to* .ii.^{e} B 747-748 meridian ... du] *om.* D 748 .xii.^{e}] 12me MP .xii.^{e} partie] moitie B du] *add.* meridian F que] *om.* P le] *om.* B li fust] fust(*interlin.*) a luy B 748-750 Et ... terre$_1$] *om.* K 749 tant] *om.* B il$_1$] *om.* M 749-750 il$_2$ ... degre] *om.* B 750 .i.] un/ung FHMNP Et ... terre$_2$] *om.* F; dont il en y a M .ccclx.] 360 MP; .iii^{c}lx. DHN 751 espaces] parties M; portions D maniere] *om.* D 753 .i.] un/ung BDFGHKMNP septentrion] *corr. to* mydi D tout] *om.* K; ou tout G 754 plus] *om.* M plus ... fust$_2$] *om.* D midi] a midi BFGHKN; vers midi M il] *add.* lui/luy ABFGHKM 754-755 en ... degre$_2$] *om.* N

1 Chap. 1, ll. 71-79.

moins eslevé d'un degré, il auroit passé de la terre .vii.^{c} estades, selonc l'aucteur de *l'Espere.*[2]

Donques convient il repliquier et multiplier cest nombre par .ccclx. Et ainsi auron que .i. circuite de la terre, a tourner tout entour, ha .ii^{c}lii. mille estades qui valent xv^{m}vii^{c}l. leues. Car une leue vault .ii. m[iles], et .i. mile vault .viii. estades, et une estade .cxxv. pas, et .i. pas vault .v. piés, et .i. pié vault .iiii. palmes, et une palme .iiii. des.[3]

Et ainsi qui pourroit environner toute la terre, et yroit chascun jour .x. leues, il auroit fait son tour en mil .v^{c}lxxv. jours, qui valent .iiii. ans, .xvi. septmaines et .ii. jours.[4]

755 .vii.^{c}] 700 MP; *corr. to* mille D   selonc] secon/secou C   l'aucteur] l'autorité K   757 il] *om.* ADFGM   cest] cellui M; ses N   nombre] nombres N   .ccclx.] 360 MP; .xxx^{c}lx. AD; .iii^{c}lx. GN; trois cens soixante HN; quatre .clx. B; *add.* fois K   758 que .i.] ou B   .i.] ung/un MP; au F; en H; ou ADG; ou/en K   entour] *add.* la terre K   .ii^{c}lii. mille] 252,000 MP; .cclii. mile AF; ii^{c}lii.^{m} H; .ii^{cc}lii. mille C; .ii.^{m} et .lx.^{60} mile D   759 .xv^{m}vii^{c}l.] 15,750 MP; .xv. mille .vii.^{c} B; .xv. mile .vii^{c}l. N   Car une leue] *om. and add. interlin.* et chacune lieue D   .ii. miles] .ii.^{m} CEFH; deux mile NP; 2 mille D   .i.] un/ung ABFHKNP   .viii.] 8 M   760 .cxxv.] 125 P; .cxxxv. F; cent et .xxv. H; .cxv. B; .c. et .xxv. N   .i .$_{1}$] un/ung ABDFGHKMNP   vault$_{1}$] *om.* ABDFGHKMN   .v.] 5 MP; cinq BD   .i.$_{2}$] un/ung ABFHKMNP   vault$_{2}$] *om.* B   .iiii.] 4 MP   761 palme] *add.* vault M   .iiii.] 4 MP   des] doie/doye HKN; doiz F   762 .x.] 10 MP   763 mil .v^{c}lxxv.] 1575 MP; .mv^{c}lxxv. A; mil .v^{cc}lxxv. C; mil .v^{c}lx. et .xv. N   .iiii.] 4 MP; quatre B   .xvi.] 16 MP   .ii.] 2 MP

2 See "Source References," line 756.

Here, Sacrobosco's measure of a degree (and of the circumference of the earth) follows Eratosthenes' estimate. (The earth's circumference along the equator is about 40,075 km; through the poles it is about 39,975 km.)

Oresme will point out in Chap. 34 that Sacrobosco later on adopts the measure of al-Farghānī (based on al-Ma'mūn [786-833 AD]) of 56⅔ miles per degree, which is also the measure that Oresme will use. See Sacrobosco, *De spera*, ed. Thorndike, p. 16 and note.

At this point we run into the problem of evaluating ancient measurements. The Greek stade is estimated at between 150 and 210 meters, which puts Oresme's value (line 760 and the appendix on "Measurements") of 204.25 m at the top end. But if we accept Eratosthenes' estimate as nearly accurate, his 252,200 stades per degree would be based on a stade equal to 159 m, or near the bottom of the range for the stade. On the other hand, using the al-Farghānī value, and Oresme's value for a [French] mile, the earth's circumference would only be 33,335 km – about 83% of the true value.

3 See the appendix on "Measurements".

4 The calculation is: 1575 days equals 4 years (1461 days, including one 29 February leap-day), plus 16 weeks (112 days) plus 2 days.

Item: pour trouver le dyamettre d'un cecle par le circuite, a une telle regle que l'en oste du cecle ou de son circuite, la .xxiie partis, et le demourant est devisé en .iii. parties, et le dyamettre du cecle est aussi grant comme la terce partie du circuite.[5]

Donques a faire selonc ceste regle, le dyamettre de la terre ha environ .v^{m} xi. leues et .vi. estades,[6] et donques a la semidyametre .ii^{m} v^{c} v. leues et .xiiii. estades.[7]
f. 15^{v} Et tant ou environ a il d'ici siques / au centre du monde.

Lequel est la plus basse chose qui soit; et qui y seroit, l'en ne pourroit de nulle partie que l'en ne montast.

Et est assavoir que l'en ne puet pas dire les mesures dessus dites precisement et a droit point. Mais il souffist et est bele chose de attaindre pres de la vraie mesure sans que il y ait notablement grant deffaute.

765 d'un] du G telle] *om.* M; *marg.* G 766 .xxii.^{e}] 22/22me DMP; .xii.^{e} HK le] du M est] soit KP 766-767 est ... circuite] prandre le tiers et par ainsi l'en aura le dyametre M 767 .iii.] 3 D; trois/troys GHNP; .iiii. F du cecle] *om.* BDG 768 selonc] secuin C 769 .ii^{m} v^{c} v.] .ii^{m} v.^{c} F 771 Lequel] qui B Lequel ... chose] Qui est le plus bas lieu M l'en] le F pourroit] *add.* aler BDFGHKN nulle] mille E 771-772 l'en ... l'en] et qu'il n'y eust nul obstacle, s'il ne se pourroit il mouvoir qu'il M 772 partie] *add.* illez(?) P 773 dessus dites] *om.* D 774 et$_{1}$] ne KM 774$_{2}$-775 et ... deffaute] bien de actaindre a ce que dessus est dit, et est moulte belle chose et joyeuse. Plus l'on doit savoir que la superficie de la terre descouverte et non descouverte est divisee en cinq parties que l'on appelle plages ou regions. *followed by lines 825-852* ("et selonc aucuns ... chapitre") *from Chap. 28, followed by the beginning of Chap. 29 (lines 858-872;* "... souvent") M 775 sans] *add.* ce GKN notablement] milleure D

5 A rough equivalent of π: 22/7 equals approximately 3, therefore the diameter of a circle is approximately 1/3 of its circumference.

6 If the earth has a circumference of 15750 leagues (ll. 758-759), the diameter would be 5013.381 leagues or 5013 leagues and approximately 6 stadia. Oresme's figure of 5011 leagues 6 stadia results from using 22/7 as the (approximate) value of π.

7 An error has been made here: half of 5011 leagues 6 stadia would be 2505 leagues 11 (not 14) stadia. It appears that someone divided the odd league of the diameter in half to get 8 stadia, and then added the entire 6 stadia of the diameter rather than 3 stadia (half of 6) to the radius.

LE .XXVII.[E] CHAPISTRE. DE LA VARIACION QUI EST POUR DIVERSES HABITACIONS

Posé que la terre fust ou soit habitee tout entour, je di premierement que ceuls qui habitent sous l'equinocial ont en tous temps les jours et les nuis equals, comme il fu dit devant.[1] Item: le soleil passe .ii. fois l'an par dessus leur testes; ce est quant il passe par l'equinocial.

Item: le soleil leur est par une moitié de l'an vers le pole artique, et par l'autre moitié devers l'autre pole, mais nous l'avons tousjours d'une part.

Item: pour ce ont il .ii. yvers en .i. an sans grant froit – c'est assavoir .i. yver quant nous avon yver, et .i. autre yver quant nous avons esté. Semblablement, il ont .ii. estés, l'un en mars ou quant nous avons ver, l'autre en septembre ou quant nous avons automptne, si comme il puet apparoir clerement.[2]

Secondement, ceuls qui habitent sous le tropique d'esté ont le soleil sur leurs testes le jour du solstice d'esté a miedi.

Tiercement, ceuls qui sont entre le tropique d'esté et le cecle artique ont les jours plus longs en esté que les nuis de tant comme il sont plus loings de

776 Le ... chapistre] *om.* HKN .xxvii.[e]] 27 DM qui est pour] de N pour] *add.* les M 776-777 Le ... habitacions] *om.* G 776-814 Le ... materiel] *om.* P 777 habitacions] *add.* .xxvii. K; *add.* .xxvii. c./ca. HN 778 Pose] posons M la ... soit] soit ou fust que la terre soit toute B fust ou] *om.* M habitee] habitable M 779 en] *om.* M temps] jours M 780 Item] Je dys aussi que M passe] *add.* par N .ii.] 2 D; deux BHM par] *om.* N 782 Item] *om.* M soleil] *add.* aussi M 783 moitie] *add.* par F moitie ... pole] moictie ils ont vers le pole antartique M l'autre pole] le polle antartique B pole] *om.* D l'avons] aultres l'avons M; les avions G 784 Item ... sans] Et pour ceste cause ont ils chascun an deux yvers sans trop M pour ce] *om.* B il] ilz DFHN ii.] 2 D; deux HN .ii. ... an] d'un an deux yvers B .i.$_1$] un/ung ADFGHNK c'est assavoir] .s. [scilicet?] D .i.$_2$] un AFGHKN 784-785 c'est ... esté] l'ung si est quant nous avons icy esté et quant nous avons yver M 785.i.] un ADFGHKN avons] *interlin.* D ont] *om.* F 786 .ii.] deux BHM l'un] *add.* est D ou$_1$] *om.* M ver] printemps M septembre] septentrion K ou$_2$] *om.* BDFM 787 autompre] autonne ABD; entomne K; amptonne GN si ... clerement] *om.* M 788 Secondement] En apres M 789 d'este] *om.* G 790 Tiercement] Et M sont] habitent GM

---

1 Chap. 5, ll. 199-201 and Chap. 18, ll. 497-498.

2 Oresme (following Sacrobosco, *De spera*, chap. 3 [pp. 104-105 and 134-135]) is noting that for those living at the equator, the sun is furthest from them when it is at either of the two tropics (the two solstices) – hence they have at those times two (not very cold) "winters". As well, the sun is nearest to them at the two equinoxes – hence they have at those times two (slightly warmer) "summers."

l'equinocial, et plus cours en yver, et ne ont onques le solei[l] sur les testes ne devers septentrion.

Quartement, ceuls qui habitent sous le cecle artique quant le soleil est au solstice d'esté – et il est au contraire de midi – il touche leur ozizon et ne resconse point. Et ainsi il n'ont point de nuit en cil jour naturel. Semblabement, quant le soleil est ou solstice d'yver, il est .i. jour naturel que il ont continuelment nuit, et que le soleil ne leur leve point.

Quintement, ceuls qui sont entre le cecle artique et le pole artique ont en esté plusieurs jours naturels qui leur sont .i. jour artificial sans nuit, et aussi en yver sont plusieurs jours naturels esquels il leur est tousjours nuit. Et tant plus s'aproche l'en du pole, tant est cil jour artificiel d'esté plus grant. Et dure en .i. lieu .i. moys, en autre .ii., en autre .iii. ou plus selonc proporcionelment est plus grande celle nuit
f. 16^{r} d'yver. Car aucuns des / signes sont tousjours sus leur ozizon, et aucuns tous temps dessous, si comme il appert par le .xxie chappitre.[3] Et tant comme le soleil est es signes de dessus, il est jour; et tant comme il est es signes de dessous, il est nuit.

792 oncques] jamais C soleil] soleis E les] leurs DKM 793 septentrion] *add.* mais tousjours de la partie de l'equinocial. M 794 Quartement ceul] Ceulx aussi M au] ou M; en HN 795 il$_2$] le soleil K 796 Et ... point$_2$] *om.* M ainsi] *add.* quant le soleil est au solstice d'esté D cil] iceluy B 797 ou] en N .i.] un/ung ABDFHKMN que il ont] *om.* D; qui est K 798 leve] *om.* D 799 Quintement] Pareillement M sont] habitent M le$_2$] *om.* G ont en esté] se habitans y a, ils ont une fois l'an que M en] *om.* D 800 plusieurs] pluseiurs E qui] *om.* M .i.] un AGHKN aussi] ainsi B yver] *add.* leur B sont$_2$] ils ont M 800-801 qui ... naturels] *om.* F 802 jour] *om.* B en .i. lieu] *om.* D .i.$_1$] un ABFGHKN .i.$_2$] un ABHNK 802-805 en ... chappitre] cellui jour en d'aucuns lieux par une sepmaine, en d'aultres lieux par ung moys, en d'aultres par lusieurs moys, et tellement que ceulx qui habitent droictement soubz le pole, leur an est d'ung jour artifficiel et d'une nuyt; et est depuys que le soleil entre en Aries et passe par Cancer jusques a la fin de Gemini. Et si ont demy-an jour sans avoir nuyt. Et aussi ont demy-an nuyt sans avoir jour, et dure depuis que le soleil entre en Libra et passe par Capriorne jusques a la fin de Pisces. Et la cause si est car leur orizon si est l'equinoctial. M 803 .ii.] deux BHN .iii.] trois BGHN ou plus] *om.* B selonc] selon ce ABGK; secon/secou ce C; aussi en yver D plus$_2$] le plus B celle] *om.* B 804 aucuns des signes sont] autun des signes est K tous] *om.* D; *interlin.* C tous temps] tousjours HKN 805 si] *om.* B .xxi.^{e}] 21^{e} D comme] que M le soleil] *om.* M 806 dessus il est] septemtrionals ils ont tousjours M jour ... est$_3$] *om.* D comme il] qu'il B; que le soleil M es signes] *om.* K dessous il est] meridionalz, ils ont tousjours M

3 Chap. 21, ll. 592-593.

Sextement, ceuls qui habitent droit sous le pole ont la moitié de l'an le soleil seur leur ozizon et continuel jour, et l'autre moitié continuelment nuit. Et ainsi qui appelle jour tout le temps continual car le soleil est sus l'ozizon, ou de soleil levant siques au soleil couchant, ils ne ont en tout l'an que une nuit et .i. jour. En la maniere que dit est de ceste moitié de la terre qui est devers le pole artique, il est a entendre de l'autre moitié, et de ses habitacions, devers le pole antartique.

Et tout cest chapistre puet apparoir legierement sans autre probation par example en l'espere materiel.

807-810 Sextement ... jour] *om.* M 807 droit] *om.* D le$_2$] du D 808 continuel] continuelment BK 809 continual] *om.* B car] que ABDHKN; *corr. to* que C 809-810 ou ... couchant] sellon et B 810 siques au] jusques a/jusqu'a HKN .i.] un/ung ABDFHKN 810 -811 En$_2$ ... est$_1$] Et tout ainsi qu'il est dit M 811 est$_1$] *om* G ceste] la M de$_2$ ... a] ainsi est il B devers ... a] de l'equinoctial au pole artique, tout ainsi doibt on M est$_3$] *interlin.* D 812 et ... habitacions] *om.* M ses habitacions] ceuly qui habitent B; *add.* qui est HKN devers le] qui de l'equinoctial au M antartique] antratique E 813-814 Et ... example] comme ung chascun peult facilement congnoistre M

LE .XXVIIIE CHAPISTRE. DE LA DIVISION DE LA TERRE EN .V. PARTIES

Si comme il appert par le .xvii. chapistre,[1] l'espere du ciel est devisee en .v. parties par les .iiii. mendres cecles; et ces parties sont appellees .v. "zones". La premiere est entour le pole artique, c'est assavoir entre le pole artique et le cecle artique.

La seconde est entre le cecle artique et le tropique de esté.

La tierce entre le tropique d'esté et le tropique d'yver.

La quarte, entre le tropique d'yver et le cecle antartique.

La quinte, entre le cecle antartique et le pole antartique. Et est assavoir que selonc ces .v. zones sont en la terre .v. parties ou plagez ou regions desquelles chascune est sous aucune de ces .v. zones du ciel dessus dites.

De ces .v. plages ou regions de terre, selonc ceste consideracion et selonc aucune, la premiere et la derreniere sont trop loing du soleil, et sont inhabitables[2] pour trop grant froit. La tierce et moienne est trop prez du soleil et sous la voie de soleil, sous la tierce zone qui est appellee "zone torride" ou "brulee", et est inhabitable pour trop grant chaut. Les autres .ii., c'est assavoir la seconde et la

815 Le ... chapistre] *om.* HKN Le ... parties] *om.* G .xxviii.^{e}] 28me MP; 26 D[3] terre] *add.* .xx[*cut off*] N en .v. parties] *om.* N .v.] 5 DP; cinq BM parties] *add.* .xxviii. K; *add.* .xxviii. chap. H 816 .xvii.] 17me MP ciel] monde N .v.] 5 P; cinq BMN 817 .iiii.] 4 DP; quatre B ces parties] cestz cinq cercles B .v.] *om.* M; 5 DP; 6 B 818 c'est ... artique$_{1}$] *om.* HN entre] entour K 820 tierce] 3^{e} D; *add.* est HN 821 quarte] 4^{e} D; *add.* est N antartique] antratique E 822 quinte] 5^{e} D; *add.* est N; *add.* si est M 823 selonc] secon/secou C .v.$_{1,2}$] 5 DMP; cinq B 823-825 ou ... consideracion] Et comme j'ay dit cy-devant ceulx qui habitent ou millieu de la tierce zone, c'est assavoir dessoubz l'equinoctial, ont tousjours les nuyts egales aux jours, ainsi le souleil ne demeure sus leur orizon chascun jour que par .12. heures, parquoy leur region ne peult pas estre trop chaulde. *followed by ll. 873-885* ("Car une cause ... encores apres") *from Chap. 29* M 824 sous] om. N .v.] 5 DP; cinq BN 825 .v.] *om.* D; cinq B selonc$_{1,2}$] secon/secou C 825-872 et selonc ... souvent] *found at the end of Chap. 26 (l. 774)* C 826 la$_{1}$] *ms* E *begins a new paragraph* 826-827 trop ... froit] inhabitables pour la grant froidure qui y est, car elles sont trop loing du soleil M 827 tierce] 3^{e} D et$_{1}$] *add.* la M sous] *om.* F 827-828 et$_{2}$ ... sous] car elles sont soubz la voye d'icellui et soubz M 828 tierce] 3^{e} D 829 .ii.] *om.* HN; deux BMP seconde] .ii./.ii.^{e} FK; deuxieme M

1 Chap. 17, ll. 451 ff.

2 In reading this text it is important for English speakers to remember that "inhabitable" in French – and Middle French – means "uninhabitable".

3 Up to this point, ms D has written "8" every time "eight" (or a "eighteen" or "twenty-eight" etc.) is meant (e.g., lines 142, 207, 254, 441 etc.). But from this point on, for the most part, D writes "6" instead of "8" as indicated in the apparatus. (See, for example, lines 982, 988, 99t, 996.) D still continues writing "6" for "six", e.g., line 1012.

quarte, ne sont pas trop pres du soleil ne trop loing, et seroient habitables et attrempees en chaut et en froit, selonc ceste consideration, et se il n'y avoit autre empeechement.

En ceste division, je ay appellé la "premiere zone" celle devers le pole artique qui tousjours nous appert; mais, selonc Aristote,[4] le pole antartique est le dessus du monde, car la partie de la terre qui est habitable est aussi comme le visage et est le devant de la terre et du monde.

f. 16v Donques est la partie du ciel / qu'elle regarde le devant du ciel, c'est assavoir la partie vers midi, du milieu de terre habitable.

Item: l'orient du premier ozizon de terre habitable est la dextre partie, car de la commence le mouvement du jour artificiel en terre habitable.

Or est il ainsi que se .i. homme avoit la face et le devant contre midi, et le dos vers terre, et la dextre main vers orient, il couvendroit que il eust la senestre main vers occident, et la teste vers le pole antartique, et les piés vers le pole artique. Et ainsi est orient la dextre partie du ciel ou du monde, et le pole antartique est le desus, midi devant, occident senestre, le pole artique dessous, mienuit desriere.

Et selonc ceste consideration, nous Latins sommes en la basse et senestre partie du monde, car nous habitons vers occident et vers le pole artique. Mais se nous considerons le propre mouvement qui est selonc le zodiaque, qui est aussi comme

830 quarte] .iiii. F; quatriesme M   et$_1$] *add.* sont et N   seroit ... et$_2$] sont B   habitables et] *om.* M   831 il] *om.* BN   avoit] *add.* aucun K   833 En] Et selon BDFGM   ay] *om.* B; *add.* parle et dit et HN   devers le] du B   834 selonc] secon/secou C   le$_1$ ... est] c'est M   835 car] que D   aussi] ainsi M   est$_3$] *om.* MN   835-836 la terre ... de] *rep.* D   836 monde] ciel K   837 Donques] Donques en la p Doncques B   qu'elle] qui D   du ciel] *om.* B   838 la partie] *om.* P   de] *add.* la ABDFHKMN   839 Item] *om.* M   de$_1$] *add.* la K   la$_1$] a la B   car] qui D   840 le] *add.* premier HN   mouvement du] *om.* M   en] *add.* la HKN   841 .i.] un/ung ABDFGHKMNP   devant] *add.* vers midi ou D   842 vers$_1$] contre D   il$_2$] *add.* en K   843 et$_2$ ... artique] *om.* FMN   844 est$_2$] *om.* BDFMN   845 midi] *add.* le M   occident] *add.* a B   dessous] *add.* la M   mienuit] *add.* le M   846 selonc] secon/secou C   848 qui est$_1$] *om.* F   selonc] secon/secou C   aussi] ainsi M

4 See "Source References," line 834.

The reasoning here is different from ours, as outlined in the following paragraphs (ll. 839-845). Aristotle considers true motion to be from right to left, which for the sun is from east to west. Thus if one lies on his back with his right arm to the east and his left arm to the west, then his head will be toward the south and his feet toward the north. Thus we should consider the antarctic pole to be the "top" of the earth, and the arctic pole to be the "bottom" and "below".

au contraire, si comme il appert par le .vi.^{e} chapistre,[5] le pole artique est dessus, et occident est dextre, et midi [devant], tousjours. Et par aventure, est cest mouvement la plus noble ou les esperes, qui ainsi sont meues, sont les plus nobles.

Item: plus grant multitude et plus beles estoiles sont en la moitié du ciel par devers le pole artique que en l'autre moitié qui est par dela l'equinocial, au moins quant a la partie du ciel qui nous appert.

Et selonc ce, nous serion en la partie du monde de dessus et a dextre et en la plus noble. Et seroit la premiere zone entour le pole artique si comme je les ordenai au commencement de cest chapistre.

849 .vi.^{e}] vii^{e} E; 6me P; lix^{e}(?) M 850 est$_{1}$] ē E devant] vant/vaut EC; vanlt P; *add.* et orient senestre M 851 ou] *add.* par aventure B ou les] antique(?) D qui] *om.* B sont$_{1}$] *om.* B meues] meuvent M 852 Item] Aussi M sont] *add.* en plus nobles sont B ciel] *add.* qui nous appert M 853 en] a ans B 854 la] *add.* plus grand D 855 selonc] secon/secou C partie] *add.* du ciel ou B de] *om.* N a] *om.* P 856 noble] *add.* partie N premiere zone] plus noble zone et la premiere B les] láy M ordenai] ordene M; ordonne K 857 cest] *om.* G

5 Chap. 7, ll. 237-239 and ff.

LE .XXIX.E CHAPISTRE. DE LA TIERCE PLAGE EN ESPECIAL

Aucuns dient que la tierce plage qui est sous la voie du soleil entre .ii. tropiques, est inhabitable, comme dit est.[1] Mais les autres dient que ce est tres noble et tres attrempee habitacion, especiaument vers le milieu sous l'equinocial, et ce fu le opynion de Avicenne.[2] Et ceuls qui maintiennent ceci, arguent ainsi: car se elle estoit inhabitable, ce seroit pour trop grant chaleur, comme dit est. Mais il n'est pas ainsi: premierement car une cause de grant chaut en nostre habitacion est ce que esté est long, et pouir ce est il plus chaut vers la fin. Et en celle region esté est bien court, car il ont .ii. estés en l'an, si comme il appert par le .xxvi[i].e chapitre.[3]

Item: une autre cause de grant chaut, quant a nous, est la longueur des jours f. 17r d'esté. Et pour ce / fait il communelment plus grant chault aprés midi que devant. Et sous l'equinocial il ont en tous temps equinoce, comme dit est ou .xxvi[i].e chapistre,[4] et ne ont pas en esté les jours si longs comme nous.

Item: la longueur de la nuit attrempe moult le chaut qui a esté de jour, et il ont en tous temps les nuis aussi longues comme les jours, comme j'ay dit souvent.

Item: une cause de grant chaut en esté est la demeure du soleil environ notre cenith pour obliquité de l'arc journal. Et l'arc journal sous l'equinocial n'est pas

858 Le ... chapistre] *om.* HK Le ... especial] *om.* DG .xxix.e] 29 M especial] *add.* .xxix. K; *add.* .xxix. chap. H; *add.* .xx[*cut off*] N 859 .ii.] deux DHMP; les deux B 860 inhabitable] habitable F que ce] qu'elle M tres$_2$] *om.* G 861 ce] se K ce fu] c'est D 862 de Avicenne] d'Avicene H; d'Aviscenne G; d'Avisene K car] que DHN 863 estoit] est K ce seroit] c'est K pour] par M 864 premierement] *om.* M car] que D est] *add.* pour B 865 est$_1$] *om.* K pour ce est] *rep.* M vers] envers C en] *om.* G celle] telle M este] comme dessoubz l'equinoctial, l'este M 866 car] que D .ii.] deux BDHMNP .xxvii.e] .xxvi./.xxvi.e ABCEGHKN; 26/26me DMP; xxvi F 867 Item] Aussi M quant a nous] si M est] *om.* B 868 communelment] communement HK grant] *om.* K 869 en] *om.* ABDFGHKMN tous temps] tousjours BCDK .xxvii.e] K; .xxvi./.xxvi.e ABCEGHN; 26/26me MP; .xxvi. F 871 chaut] longueur D de$_2$] le HKN 872 en] *om.* B nuis] jours D aussi] ainsi N jours] nuis D comme ... souvent] *om.* D 873 Item] Car M une] *add.* autre ADFG est] *om.* DK; temps si est M la] *add.* longue M notre] le M 874 pour ... sous] qui se fait pour l'obliquité de l'arc journal en l'espere oblique. Mais l'arc journal en l'espere oblique, droicte dessoubz M 874-876 pour ... cenith] *om.* K

1 Chap. 28, ll. 827-829.
2 See "Source References," line 862.
3 This issue is covered in Chap. 27 (not 26). See ll. 785-787.
4 This issue is also covered in Chap. 27 (not 26). See ll. 790-793.

oblique, mais est droit, si que le soleil est tantost passé et n'arreste gueres entour le cenith. Et ceci appert par example en l'espere material.

Item: tout le ciel leur appert et toutes leur estoiles qui sont vers nostre pole et vers l'autre. Si dient aucuns que il a vers l'autre pole aucunes estoiles que nous ne po[v]ons veoir, qui sont de froide vertu par lesquelles est attrempee la chaleur es regions qui sont dessous l'equinocial.

Item: aucuns dient que en celle plage vers orient est paradis terrestre en une montaigne.

Item: il dient que pres du tropique de esté par de sa et dessous a trop grant chaleur desatrempee et terre inhabitable et mal [habitable], et aussi sous le tropique d'yver et emprés par dela; et de ceci parlerai je encor aprés.[5]

875 mais est droit] *om.* M si que] siques F; si parquoy M est ... passe] passe tantost oultre M n'arreste] ne demeure M le_2] leur ABCDFGHMN 876 en] de D 877 Item] Aussi M leur appert] *om.* K $leur_2$] les DHKN nostre] le DM 877-878 et vers l'autre] artique et celles qui sont vers le pole antartique M 878 Si ... aucunes] Et disent aulcuns que vers cellui pole antartique a aulcunes M aucunes] autres HN 879 povons] poons E de] *om.* B vertu] *om.* B lesquelles] la froidour(?) de clle B es] des B 881 Item ... dient] Et disent encores aucuns M plage] *add.* ou zone M terrestre] terre C 883 Item] *om.* D Item il dient] Plus disent M par ... dessous] *om.* M 884 desatrempee] *om.* B; desattrempeement ACFGHKMN inhabitable et] *om.* B et_2] ou K et_2 ... habitable] ou de mauvaise habitacion M habitable] habitabla/habitablei E; habilitable P 885 empres] apres N; ung petit M par dela] *om.* B; *add.* pour l'excessive chaleur M parlerai je] parleroy B je] *om.* K

5 Chap. 31, ll. 944 ff., and Chap. 40, ll. 1259-1293.

LE .XXXE CHAPISTRE. DE LA QUARTE PLAGE EN ESPECIAL

La quarte plage est entre le tropique d'yver et le cecle antartique, et selonc la consideration du .xxviie chapistre[1] et l'opinion d'aucuns, celle plage est aussi attrempee et aussi bien habitable comme ceste seconde ou nous sommes.

Et pour ce dient il que il y a gens et royaumes et habitations tout aussi comme de ceste part, et ont yver quant nous avons esté, et esté quant nous avons yver, et automne quant nous ver, et ver quant nous autompne. Mais nuls ne puet aler de ci la, ne venir de la ici, ne par mer ne par terre, si comme il dient, pour ce que il couvendroit passer par la tierce zone, dont j'ay parlé au chapistre precedent, qui est inhabitable selonc ceste consideration, et mesmement sous les tropiques, comme dit est.

Et dient que illeuques sont ceulz que l'en appelle "antipodes", c'est a dire gens qui ont leurs piés contre les nos, et pour ce que il sont a l'opposite partie de la terre, aussi comme se il fussent sous nous, et nous sous eulx.

Ceste opinion n'est pas a tenir, et n'est pas bien concordable a nostre foy, car
f. 17^{v} la Loy Jhesucrist a esté preechice par toute terre / habitable. Et selonc ceste opinion, celles gens ne n'auroient onques oy parler ne ne pourroient estre subjecs a l'eglise de Romme. Pour ce repreuve Saint Augustin ceste erreur en ceste opinion, li[br]o .xvi.^{o} *De civitate Dei.*[2]

886 Le ... chapistre] *om.* HKN Le ...especial] *om.* G .xxx.^{e}] 30/30me DMP quarte] 4 D; .iiii. H especial] *add.* .xxx. K; *add.* .xxx. ca. HN 887 selonc] secon/secou C 888 .xxvii.^{e}] 27/27me DMP; .xxvi.^{e} HKN et] *add.* selon M aussi] ainsi M; *add.* bien B 889 et ... habitable] *om.* B 890 y] *om.* B et$_{2}$] aux B tout aussi] *om.* M 891 ceste part] *illeg.* D avons$_{1}$] *om.* K; avon E et$_{2}$ *add.* ils ont P avons$_{2}$] *om.* N; avon E 892 autompne$_{1}$] auptonne N; emptonne GK; aussi B nous$_{1}$] *add.* avons ADFM ver$_{1}$ ... autompne] avons ver ils ont automne et econ̄ B nous$_{2}$] *add.* avons ADFK autompne$_{2}$] amptonne G; auptonne N; emptonne K peut] *add.* y B 893 la$_{1}$] *om.* B venir] *om.* BDFGM ne$_{2}$] *om.* M 894 convendroit] convient K tierce] 3 D zone] sone K 895 selonc] secon/secou C 897 ceulz ... appelle] *om.* HKN l'en] on D 898 nos] nostres KM 899 aussi] ainsi M se] *om.* B 900 et] ne B concordable] accordable BM 901 Loy] foy D; foy de M Jhesucrist] *illeg.* D; Ih'ucrist E; Ihucrist AF; Ihūcrist/Jhūcrist BCGHKM; Ihesucrist NP selonc] secon/secou C 902 celles] telles K; toutes nacions de M ne$_{1}$] *om.* BGM oncques] pas M parler] *add.* de Jhesucrist M 903 en] ou HKN en ... opinion] *om.* M 903-904 libro .xvi.^{o}] li.^{o} .xvi.^{o} CGE; li. .xvi.^{o} F; li. 16 M; 51° 16° P; lib^{o} 16 D; lib^{o} .xvi.^{o} N

1 Chap. 27, ll. 790-793 and ll. 810-812.

2 See "Source References," line 904. Some scribes have mistaken "li^{o}" (i.e., "libro") as roman numerals for "51".

Item: ce ne semble que .i. divinement dire que il soient tant de peuples et de royaumes sans ce que l'en en ait par de sa ne ne puisse avoir aucune experience, mais dire le tant seulement pour .i. arguement d'ymagination. Et se tier arguemens concluent, il s'ensuirroit que toute la terre fust couverte d'yaue, aussi comme l'yaue est couverte de l'air, et l'air de feu, et ainsi des autres esperes du monde, si comme il fu dit au premier chapistre, et au secont.[3]

Et pour ce devroit l'en miex dire que celle quarte plage est couverte d'yaue ou de mer en la plus grant partie, et que il ne a de la terre descouvert fors ce de quoy l'en puet avoir experience.

Item: [je] vueil monstrer par raison que il n'est pas semblable quant a estre habitee de ceste seconde plage et de celle quarte, pourquoy il est asavoir que il sont .ii. principauls causes de plus grant chaleur en une region.

La premiere est quant le soleil est plus eslevé sus l'ozizon, et vient plus pres du cenith. Car lors il giete ses rays plus droit sur la terre, et fait plus droite reflexion de lumiere qui est cause de chaleur.

La seconde cause est quant le soleil est plus pres de terre ou de nos testes, et moins de distance; et ce est quant il est en *oppositum augis*, comme il appert par le

905 Item ... .i.] Aussi me semble que c'est ung M .i.] un/ung ABDFGHKNP soient] soit D peuples] pueples E peuples et de] *om.* B de$_2$] *add.* tant de P 906 ce] *om.* B en$_2$] *om.* GN en ait ... ne ne] *om.* M; par de ca ne D en ait ... avoir] n'y aille par mer ou par terre ou B ait] *add.* ou puisse avoir K ne ne ... avoir] *om.* K 907 dire le] que le dire K le] *om.* G; lay HN; loy B seulement] *om.* D pour] par KM .i.] un/ung ABDFKMN d'ymagination] *add.* est chose frivols M tier] cieulx BN; telx D 907-919 Et ... secont] *om.* M 908 il] *rep.* E 908-909 d'yaue ... couverte] *om.* D 909 de l'air] d'air K; de air B de$_2$] du F ainsi] aussi D si] *om.* HK 910 secont] .ii. F 911 Et ... dire] Mais est bien de croyre M dire] *om.* B quarte] 4^{e} D est] *om.* D; *add.* toute M 912 de mer en] au moins pour M partie] part M ne] n'y DN la$_2$] *om.* KN 914 je] *om.* EP 914-916 monstrer ... de] aussi demonstrer que la disposicion d'icelle quarte plage ne peult estre egalement eschauffee ou temperee comme la deuxieme en laquelle nous sommes, car en icelle quarte plage y a une cause principale pour laquelle il y a M 915 pourquoy] *corr. from* de quoy A 916 .ii.] 2 D; deux BHNP une] leur K une region] icelle plage que en la seconde laquelle M 917 La ... est] Si est car M plus$_1$] *om.* M vient ... du] approche le M pres du] vers le B 918 Car] que D Car lors] *om.* M giete] ierte A la] *om.* M 919 de] *add.* la G 920 seconde] 2 D; .ii. F et] et a/ha BCFGH 920-923 La ... car] Et avecques ce ils ont le soleil plus pres de terre, qui est aussi cause augmentative de chaleur pour laquelle icelle plage est plus eschauffee que la seconde. La raison pourquoy le soleil est enleur esté plu pres de terre, c'est car le soleil est in oppositum augis, et en nostre esté, nous l'avons in augis mesmement quant il est ou tropique d'esté ou environ, et ainsi il est plus loing de terre en l'esté de la deuxieme plage qu'il n'est pas en l'esté de la quarte plage et M

3 Chap. 1, ll. 79-107 and Chap. 2, ll. 109-124.

.xxie chapistre.[4] Et adonques eschauffe il plus fort se les autres choses estoient pareilles, car, si comme dit Alkindus ou *Livre des impressions*,[5] la chaleur des planetes est plus forte de tant comme il sont plus pres de l'air.

Or est il ainsi que le soleil est en *aux solis* et environ en nostre esté, et donques est la premiere cause de chaleur forte, et la seconde, foible. Et pour ce, avons nous en esté chaleur attrempee souffisaument. Et en yver le soleil est pres de la terre et de nous, et en *oppositum augis*, si comme il appert ou .xxiie chapistre,[6] et ainsi est la premiere cause de chaleur foible, et la seconde, forte. Et pour ce avons nous froit attrempé souffisaument. Mes ceulz qui habiteroient en la quarte plage auroient autrement. Car en leur esté, quant nous avons yver, les .ii. causes de chaleur sont fortes, et est le soleil pres de terre et haut sur l'ozizon, si a aveques chaleur desordenee; f. 18^{r} et / en leur yver, quant nous avons esté, l'une et l'autre cause de chaleur est foible, et y a desordeneement grant froit si que par ceste raison celle quarte plage est inhabitable ou plus mal habitable que la nostre. Et se *aux solis* est meu chascuns .c. ans .i. degré ou environ, si comme aucuns dient, il sera aucune fois au contraire se le monde duroit tant. Et seroit par ceste raison ceste partie et ceste plage desatrempee en chaut et en froit ou mal habitable; et l'autre, c'est asavoir la quarte plage, seroit

922 .xxie] 21/21me DP fort] *add.* que F; *add.* que autre fois DG; *add.* que une autre fois que B 923 pareilles] parelles F Alkindus] Abbendus D; Alkoindus/Alkomdus P la] que la M 924 est]et N est ... l'air] de tant qu'elles sont plus pres de terre, de tant est plus forte leur influence M il] elle F plus$_2$] *om.* K plus pres] p's B 925 et$_1$] ou BK este] plage B 925-935 Or ... meu] Item, quant nous avons yver, le soleil qui est adonc pres de terre mitigne et atrempe la froidure. Et quant ilz ont yver en la quarte plage, le soleil est adonc loing de terre; parquoy en icelle zone est causes desordonnee froidure. Ainsi il appert que icelle plage n'est pas de egale disposicion a la deuxieme, parquoy ne peult estre peuplee ne habitee comme ceste. Aux solis, de quoy est dessus parlé, se meut M 926 seconde] 2e D; .iime P 927 attrempee] *add.* suffisant et B 928 augis] *corr. from* auguis E ou] par le FN .xxiie] 22/22me DP; .xiie GHKN ainsi] adoncque B 929 seconde] 2 D; .iime P 930 habiteroient] habitent K auroient] ont B 931 .ii.] 2 D; deux HKNP 932 aveques] illec HN; illecques BF; illouc D 934 a] *om.* D froit] fort N quarte] 4 D 935 que] *add.* n'est G .c.] 100 DMP; cent HN 936 .i.] un/ung ABFGHKNP .i. ... environ] d'ung degré d'occident en orient ou environ ung degré M il] Et M aucune] autre HN; une M au] *om.* D contraire] *add.* point M 937 seroit] ha B ceste partie] *om.* B. plage] *add.* mal habitable et B 938 ou] et M ou mal habitable] *om.* B seroit] *om.* B 938-939 l'autre ... et] *om.* D

4 Chap. 21: this appears to be an error. The reference should be to Chap. 22, ll. 641-642.

5 See "Source References," line 923.

6 Chap. 22, ll. 642-644.

habitable et attrempee se elle est descouverte d'yaue. Mais encor ne scet l'en pas bien comme *aux solis* est meu, si que ceste chose ne ceste consideracion n'est pas de tout certaine.

939 et] *add.* bien M est] estoit ADFGHKMN; est~~oit~~ C; seroit B Mais] *om.* H ne scet] ne soit AB; n'estoit HN 940 comme] qui est D; si G si que] siques FM; sic que B chose] *om.* D chose ne ceste] *om.* M de tout] du tout comme dit est B

Addenda: Chapter 30

Oresme argues that there are two causes of excessive heat in a region: (1) the sun is higher in the sky and the rays more direct; (2) the sun is closer to the earth (at the *oppositum augis*).

In our (northern) summers the sun is higher in the sky but nearer the *aux solis* (further from the earth); therefore it is strong for the first reason and weak for the second. In our (northern) winters the sun is lower but nearer the *oppositum augis*; therefore the sun is weak for the first reason and strong for the second. In both seasons the strength or weakness of the first reason is countered somewhat by the weakness or strength of the second reason.

However for the (southern) summers – the northern winters – the sun is higher (and stronger) and nearer the *oppositum augis* and therefore also stronger. For the (southern) winters – the northern summers – the sun is lower and weaker and nearer the *aux solis* (further from the earth) and weaker.

So the summers in the south are hotter than our summers in the north, and the winters in the south are colder than our winters in the north.

LE .XXXIE CHAPISTRE. DE LA DIVISION DE L'ABITACION DE LA TERRE SELONC AUCUNS

L'en doit ymaginer un cecle en la superficie de la terre sous l'equinocial, et .i. autre cecle qui interseque cesti ci en [c]rois, et qui passe par orient et par occident de terre habitable. Ces .ii. cecles ici devisent la terre en .iiii. quartiers [e]quals desquels le quartier ou nous sommes tant seulement est habitable, et non pas tout. Car il en a aucune partie vers l'equinocial qui n'est pas bien habitable pour cause de trop chaut, et une autre vers le pole artique mal habitable ou inhabitable pour trop froit. Par quoy il appert que la terre habitable n'est pas roonde, ne a maniere de cecle – et ce dit Aristote[1] – mes est comment le quart de la superficie d'une espere, lequel quart est retrenchié as .ii. bous, ou comme un escu large coffin en concave qui est retrenchié au bout dessous; et ne puet pas la mappemonde estre bien proprement figuree en chose plaine, mais tant seulement en chose boucue a maniere d'une partie d'espere. Et ha la terre habitable .iiii. cuygnes qui sont appelés les ".iiii. angles de terre habitable". Li un puet estre appelé "l'angle d'orient d'esté"; l'autre, "l'angle d'orient d'yver"; l'autre, "[l']angle d'occident d'esté"; l'autre "l'angle d'occident d'yver".

955 cuygnes] *read* coin

942 Le ... chapistre] *om.* HKN .xxxie] 31/31me DMP de] et D 943 l'abitacion de] *om.* G de la terre] *om.* B terre ... aucuns] *om.* (*cut off?*) N; *add.* .xxxi. c. N selonc] secun C selonc aucuns] *om.* DH; *add.* .xxxi. chap. H aucuns] l'oppinion d'aucuns G; les anciens .xxxi. K 944 un] .i. DG de la terre] *om.* D sous] selon D .i.] *om.* D; un/ung AFGHKMNP; *corr. from* en B 945 crois] trois ECPGN 946 habitable] inhabitable D; *add.* et par les deux poles M .ii.] deux BDHMP ici] *om.* D .iiii.] 4 P; quatre BMN quartier] quarte K equals] quals E desquels] desquienx N 947 pas] *add.* encor le M 948 habitable] habitee AFGM de] de la M chaut] grant chaut ABDFG; grant chaleur M 949 autre] *add.* partie M mal ... ou] qui est M ou] et B ou inhabitable] *marg.* C trop] *add.* grant B 949-950 trop ... a] la trop grant froideur. Et n'est pas cestuiquartier ou icelles partie en quoy nous sommes reonde ne en M 950 terre] *om.* B a] n'a B; en H et ce] comme M 951 est comment] comme BDFGM; est comme AKN 952 retranchie$_1$] transche B as .ii.] en deux B .ii.] deux HKMNP un] .i. G en] ou BHKN 953 au] ou ABFGHN pas] *om.* H bien] *om.* K proprement] *om.* B 954 a$_1$] en HMN Et] *add.* ainsi M a$_2$] est B 955 habitable$_1$] *add.* divisie en B .iiii.$_1$] 4 P; quatre ABDHKM .iiii.$_2$] 4 MP; quatre BD 956 l'angle$_1$] l'anglet K 956-957 l'autre ... este] *om.* D 957 l'angle$_1$] angle E

1 See "Source References," line 951.

f. 18^v / Et est asavoir que on puet entendre par orient chascune partie la ou soleil lieve sus aucun ozizon, et ainsi est il partout orient ou regart d'aucun ozizon et partout occident. Mais, orient puet estre pris autrement pour la premiere partie de terre habitable vers soleil levant, ou pour l'orient du premier ozizon de terre habitable vers soleil levant; et occident aussi pour cellui du desrenier ozizon vers soleil couchant. Et ainsi sont il pris en commencement de cest chapistre.

En retournant a propos, la porcion dessus dite qui est terre habitable, puet estre devisee en .iii. manieres; en une maniere selonc les hystoriens que l'en appele "cosmografes", comme sont Plinius, Pomponius, Solinus, Prescian, Anselme[2] et plusieurs autres; et aucuns astrologiens, si comme Albategni[3] qui la divisent en .iii. parties – c'est asavoir Asye vers orient, Affrique et Europe vers occident; et est Affrique devers l'equinocial, et Europe devers septentrion. Et entre ces .ii. parties est la Mer de Myterre.[4] Et devisent les aucteurs chascune de ces parties en plusieurs royaumes et regions, mais ceci appartient a la mappemo[n]de.

958 est] *om.* K assavoir] a entendre B on] l'en B; l'on M entendre] attendre N la] *om.* D ou] ou le ABDFHKMN 959 aucun$_1$] autre H; l'autre N regart] regaent K d'aucun$_2$] d'autre N et$_2$] *add.* aussi est il M 960 de] de la D 961 vers$_1$] vers le KN 961-962 ou ... levant] *om.* BFKM; *add.* ou pour l'orient du premier D 962 aussi ... ozizon] vers la derrenier partie de terre habitable vers occident ou M vers] du G 963 ainsi] aussi D en] au PBDFKM; on G; ou AH 965 en .iii. manieres] *om.* M .iii.] trois/troys BP; deux HN; quart D .iii. ... en$_2$] *om.* F en une maniere] *om.* B selonc] secon/secou C hystoriens] histoires M que l'en] con K 966 comme sont] *om.* D Pomponius] *om.* K Prescian] Priscan K; Priscian HN; P'scian ABDFG Anselme] Aussellme HN 967 et ... astrologiens] *om.* G Albategni] Abbategni H; Albatogni G .iii.] 3 DP; trois HM 968 asavoir] *add.* en KN; *add.* vers G Asye] *marg.* A; Aise K vers$_1$] droit P orient] *add.* en HKN et$_1$] *add.* en K Europe] Eusope K 968-969 et$_2$ ... devers$_1$] mais Affrique est vers M 969 devers$_1$] vers K Europe] Eusope K .ii.] 2 D; deux BHMNP parties] *om.* N 970 plusieurs] *add.* parts et B 971 et regions] *om.* B mappemonde] mappe mode E

2 See "Source References," line 966.

3 See "Source References," line 967.

4 Mediterranean Sea.

DIAGRAM: CHAPTER 31

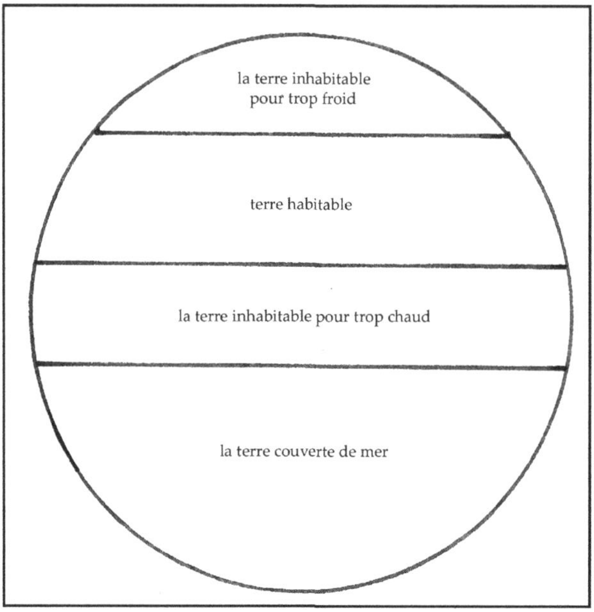

Complete diagram: E CP
Partial diagram: M[5]
Alternative diagram: BFGHKN
No diagram (or space only): D

la terre ... froid] *om.* E

5 While the diagram in ms M associated with Chap. 31 is partial, it is repeated and combined with the diagram for Chap. 32 where "circulus equinoctialis," "tropicus Cancri" and "circulus articus" are added.

DIAGRAM: CHAPTER 31 ALTERNATIVE

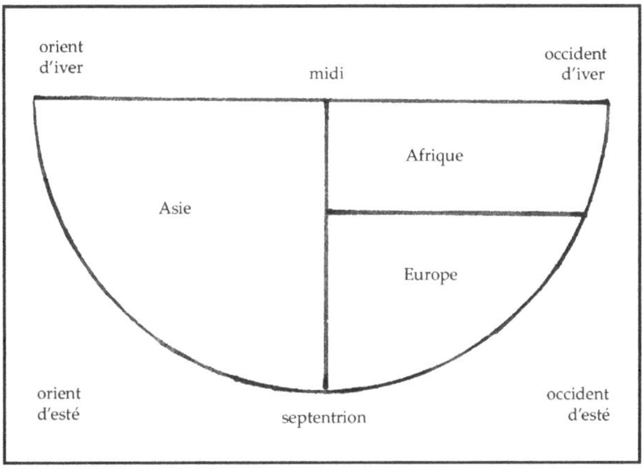

Complete diagram:[6] BFGHKN

midi] *om.* K   Asie] Assye B; Asye FN; Ayse HK

6 Note that this diagram follows the early tradition of placing "north" at the bottom.

LE .XXXIIE CHAPISTRE. DE LA DIVISION DE TERRE HABITABLE SELONC LES ASTROLOGIENS

f. 19^{r} Une autre maniere est selonc Ptholomee[1] qui devise / la terre habitable per .xxvi. lignes equidistantes de l'equinocial, et qui procedent d'orient en occident.

La premiere est sous l'equinocial la ou en tous temps le jour artificiel est de .xii. heures. Mais Ptholomee ne dit pas que il y habite gens.

La seconde passe la ou le jour ha .xii. heures et .i. quart.

La tierce, la ou il ha .xii. heures et .ii. quars, c'est .xii. heures et demie.

La quarte, la ou il ha .xii. heures et .iii. quars, et ainsi ensuiant, en acroissant le jour a chascune ligne du quart d'une heure siques a la .xxvi.^{e} ligne la ou le jour ha .xviii. heures et demie, et est – ce dit Ptholomee – en la Mendre Bretaigne – c'est par aventure, Escoce – ou c'est en aucun ills.[2]

972 Le ... chapistre] *om.* HKN .xxxii.^{e}] 32/32me DMP chapistre] *add.* est B de$_2$] *add.* la BDGH habitable] *om.* K; *add* .xxxii. N; *add.* .xxxii. chap. H 973 selonc] secun C selonc les astrologiens] *om.* HN; *add.* .xxxii. K 974 selonc] secun C Ptholomee] Ptolomee F; Tholomee C; Protholomee D; Ptholome G .xxvi.] 26 DMP 975 lignes] signes G 976 sous] dessoubz D en] *om.* D le ... est$_2$] les jours artifficielz sont M artificiel] *om.* K .xii.] 12 DMP 977 Ptholomee] Ptolomee; PTolomee C; Protholomee D il y habite] il y habitent F; la habitent M 978 seconde] .ii. F; deuxieme M; 2 D jour] *add.* plus D; *add.* plus long FGHKMN .xii.] 12 DMP .i.] *om.* D; un/ung BFGHKNP .i. quart] ¼ M 979 tierce] .iii./.iii.^{e} BF; 3me DM la$_2$] *om.* D .xii.$_1$] 12 DMP; douze B .ii.] 2 D; deux BHP; .xii. N .ii. ... demie] ½ M c'est] ce sont P c'est ... et$_2$] qui vallet B .xii.$_2$] 12 DP demie] *add.* car deux quars valent demie heure HN 980 quarte] .iiii. F; 4me M; *add.* passe P il ha] il y a D .xii.] 12 MP; douze B .iii.] 3 D; trois/troys BFGHNP .iii. quars] ¾ M; *add.* la quinte la ou il a .xiii. heures K ainsi] aussi D ensuiant] de^{r} aultre B acroissant] encroissent D; croissent BM 981 a$_1$] *om.* K a$_1$ ... siques] de ¼ de heure en chascune ligne jusques M du] d'un BK; le D d'une] de B siques] jusques BKP .xxvi.^{e}] 26me DMP ligne la] *om.* M 982 .xviii.] 18 MP; 16 D ce] comme M Ptholomee] Ptholomee C; Protholomee D; Ptolomee F; Tholomee K en] a HN 983 Escoce] Escousse B; Escosse HKM ou ... ills] *om.* M c'est en] *om.* B; ce est/en F en] *om.* DN

1 See "Source References," line 974. See also the appendix, below, on "Climes," section 1.

2 Ptolemy (*Almagest*, Book II, cap. 6.) refers to "Little Britannia" which is his name for Ireland (and not Brittany / Bretaigne). The position of this parallel (59° 30′ as calculated from the length of the longest day, 18h 30′) passes somewhere between the Orkney Islands and the Shetland Islands, well north of Ireland.

Et encore met Ptholomee habitacions a pres vers septentrion. Mais tele habitacion n'est pas continuel, si comme il note en son *Quadriparti*,[3] car comme dit Haly, il y a plusieurs lieus inhabitables pour trop froit.

Par le jour artificiel l'en entent le temps de soleil levant siques a soleil couchant comme il fu dit ou .xviii.^{e} chapistre.[4] Et par une heure je entent la .xxiiii.^{e} partie du jour naturel, car il ha .xxiiii. heures equales. Mais aucunefois prent l'en une heure pour l'espace de temps ou il lieve demi signe, et est appelee "heure naturel". Et aussi a il en chascun jour artificiel .xii. heures et autant en la nuit, et ne sont pas equales, si comme il puet apparoistre par le .xx.^{e} chapistre.[5]

La tierce maniere de deviser la terre mettent Albumazar, Alphagranus[6] et plusieurs astrologiens qui divisent terre habitable en .vii. parties appelees ".vii. climats" par .viii. lignes equidistantes de l'equinocial. La premiere ligne commence le premier climat, la seconde le secont, et ainsi en aprés, et la .viiie [ligne] fine le .viie climat.

Et est chascun climat denominé d'aucune chose notable qui est en ycelui. Le premier est appelé "clima Dyameroes" pour cause d'une isle qui y est qui a nom

984 Ptholomee] Ptolomee CF; Protholomee D   a pres] *om.* BDFGM; a A; en .iii./trois HKN   985 il] *add.* le KN   Quadriparti] Quadripertyt M; Cadriparti N   985-986 dit Haly] il dit HKN   986 lieus] pays M   pour] par M   987 l'en] on M; l'an G   temps] jour HN   de] du DM   siques] jusques DHKMP   a] au DG   988 il] *om.* K   .xviii.^{e}] 18me MP; 16 D; .xxviii.^{e} N   .xxiiii.^{e}] 24me DMP   989 il ha] il y a DK   .xxiiii.] 24 MP; 14 D   l'en] on M; l'an G   990 ou] ou quel M   991 jour] *interlin.* A   artificiel] naturel M   .xii.] 12 DM   992 si] *om.* B   peut apparoistre] appert DM   apparoistre] apparoir BGK   par le] ou G   .xx.^{e}] 20me DM; .xxi. K   993 tierce] 3^{e} D   maniere] partie HK   la terre] *om.* D   Albumazar] Abumasar HKN   Alphagranus] Alfaganus A; Alfigramus B; Alfraganus FM; Alfagranus G   994 plusieurs] *add.* autres BG   terre] la terre M   .vii.$_{1}$] 7 DP; sept HM; sep N   .vii.$_{2}$] 7 DP; sept H; les sept M   995 .viii.] 8 MP; 6 D; *corr. from* sept B   premiere] *add. and del.* climat G   commence] comme H   996 seconde] 2^{e} D   secont] 2 D; deuxieme M; *add.* climat F   et$_{1}$ ... apres] *om.* M   en] *om.* FHNP   en apres] des autres B   et$_{2}$] et en N   .viii.^{e}] 8me MP; 6 D   ligne] *om.* EACDFGMP   fine] *om.* HN; est le fin K   le] du HKN   .vii.^{e}] 7me MP; septiesme BK   998 est$_{1}$] *om.* B   climat] *add.* se B   d'aucune] par aucune BM   Le] Car le M   999 est ... clima] climat est nomine B   Dyameroes] Dyamerves A; Dyameroes GHN; Diamerros F   qui y est] *om.* B; qui est en cellui climat M

3 See "Source Reference," line 985.
4 Chap. 18, l. 482.
5 Chap. 20, ll. 552-554.
6 See "Source References," line 993.

“Meroe”, et est cil isle ou fleuve de Nil. Le secont a nom “Dyacienes” car il y a une isle qui est appelee “Cyene” qui est ou Nil, au-dessous de Meroe. Le tiers est appelé “Dialexandrios” car illeques est Alexandrie, la cité; le quart est appelé “Dyarodes”, la ou est l’isle de Rodes; le quint, “Dyaromes”, la ou est Romme; le sixte “Dyaboristenes” la ou est le fleuve Boristen; le septime “Dyaripheos” car il est ver les montaignes qui sont appelees “Ryphees”.[7]

f. 19v / Item: Albumazar[8] attribue les .vii. climats as .vii. planetes et dit que, selonc les philosophes de Perse, le premier est attribué a Saturne pour ce que Saturne y a aucune seignourie ou influence especial; le secont est attribué a Jupiter; le tiers a Mars; le quart au soleil; le quint, a Venus; le sixte a Mercure; le septime a la lune, selonc l’ordre des planetes. Mais il dit que, selonc les Grecs, le premier est attribué a Saturne, le secont au soleil, le tiers a Mercure, le quart a Jupiter, le quint a Venus, le sixte a la lune, le septime a Mars.

1001 Meroe] Merve A de] du M Nil] Nill B; Nille D Le secont] *om.* D Dyacienes] Dyatines B; Dyacienes F; Dyacenes HN; Dyacrenes K il] *om.* N 1002 qui est$_1$] *om.* BDFHKMN Cyene] Tyene B; Ciene FHK; Cyenne N qui$_2$ ... Meroe] *om.* M ou] au F; en N; *add.* fleuve de K au-dessous] dessous F; ou au-dessous N Meroe] Merve A tiers] 3 D 1002 Dialexandrios] Dyalexandrios DFGHMK car] que D Alexandrie] Dyalexandrie Hk le] *om.* N quart] 4 D est appele] *om.* ADFGHMN 1003 la$_1$] *om.* D Rodes] Rosdes N le quint] 5 D Dyaromes] Dyarommes K la$_2$] *om.* D Romme] Rome FGM; *add.* la cite G le sixte] *om.* G; le 6 D 1003-1004 sexte ... Boristen] sixiesme la ou est appelle le fleuve Briscon s’appelle Dyaboristenes M 1004 Dyaboristenes] Dyasoristenes A ou] la ou BCFGKN Boristen] Boriste N; de Soristen A; de Boriste K; de Boristen G; de Bouriston F; de Dyabor~ D septime] 7 D; .vii.^{e} BHN; *add.* est appelles N Dyaripheos] Dyarifeos BG; Dyarhipeos D car] que D 1005 les] *om.* M qui sont] *om.* M Ryphees] *om.* B; Rhipheos D; Riphees FM; Riphiees G; Riphes KN 1006 Item] *om.* M Albumazar] Albumasar FHKN .vii.$_1$] 7 DMP; sep N .vii.$_2$] 7 P; sept BDM selonc] secun C 1007 philosophes] philophes/phisophes A de Perse] *om.* M; que (*later hand*) D pour ... Saturne$_2$] et B 1008 ou] *add.* signiffiance ou M secont] 2 D est attribue] *om.* BD tiers] 3 D 1009 quart] 4 D le quint] 5 D; E *starts a new paragraph* sixte] 6 D; .vi.^{e} BHN septime] 7 D; .vii.^{e} BHKN 1010 selonc$_{1,2}$] secun C il dit que] *om.* M 1011 secont] 2 D tiers] 3 D quart] 4 D quint] 5 D 1012 sixte] 6 D; .vi.^{e} BKN septime] 7 D; .vii.^{e} BHKN

7 For meaning of these names, see the appendix on “Climes”, section 2.

8 See “Source Reference,” line 1006. The linking of the seven climes to the planets is summarized below in the appendix on “Climes”, section 3.

DIAGRAM: CHAPTER 32

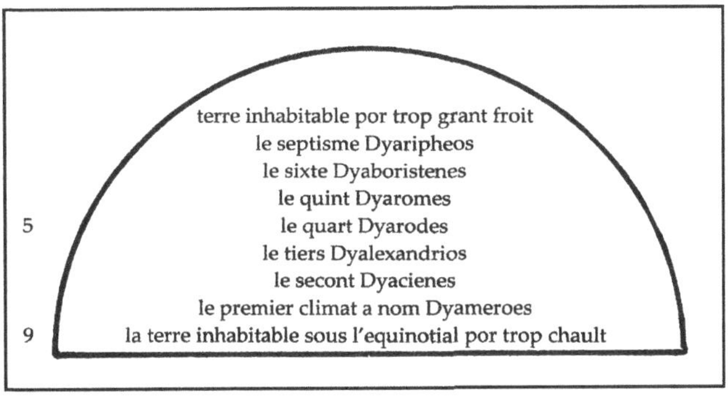

Complete diagram:[9] E BCFGHKMNP
Partial diagram: D

1 terre ... froit] *om.* BFGHKMN; Climats H 2 le septisme] *om.* H; le 7me P; clima BFGN; clima 7^{m} M; climat K Dyaripheos] Diarpheos GN; Dyarifeos BH 3 le sixte] le 6me P; clima BFGHN; clima 6 M; climat K 4 le quint] clima BFGHN; clima 5 M; climat K 5 le quart] clima BFGHN; clima 4 M; climat K 6 le tiers] clima BGFHN; clima 3 M; climat K Dyalexandrios] Dyalexadrios EC; Dyalexendries K; Lexendrios B 7 le second] clima BFGHN; clima 2 M; climat K 8 le ... nom] clima BFGHN; clima 1 M; climat K 9 la ... chault] *om.* BFGHKMN

9 In mss ECP the climes are listed from north down to the south; in mss BFGHKMN, they are listed from south down to the north.

In ms M which combines this with the diagram from Chap. 31, climes 1 and 2 are between the equator and the Tropic of Cancer, and the other 5 climes between the Tropic of Cancer and the Arctic Circle.

LE .XXXIIIE CHAPISTRE. DE LA LONGITUDE DES CLIMATS

Chascun climat ha longitude et latitude, c'est a dire longueur de lay.[1] Et est chascun climat plus long que lay. La longueur ou longitude est en procedent d'orient en occident; et latitude ou le lay en procedant, ou on alant, de midi vers septentrion.

Le premier climat, selonc les astrologiens, contient de long la moitié du circuite de la terre, lequel circuite aucuns mesurent en la maniere qui fu dicte ou .xxve chapistre,[2] c'est assavoir par estades desqueles .viii. valent un millier. Et ainsi chascun degré du circuite de la terre est .lxxxvii. milliers et demi.

Mais Alphagranus[3] et aucuns autres mesurent le circuite de la terre par milliers qui sont d'autre maniere car il dient que chascun degré du circuite de la terre ha .lvi. f. 20^{r} milliers / et .ii. tierces d'un millier tant seulement. Et ainsi auroit tout le circuite .ii.^{c} mille et .iiii.^{c} milliers – ce sont .c.^{m} et .ii.^{c} lieues – et la moitié du circuite .c.^{m} et .ii.^{c}

1013 Le ... chapistre] *om.* HKN .xxxiiie] 33me DMP climats] climas A; *add.* .xxxiii. KN; *add* .xxxiiie chap. H 1014 longitude ... dire] *om.* K et latitude] *om.* BD; *marg.* A longueur] long K de] et ABCDFGHKMN lay] largeur N; latytur M est] *interlin.* A 1015 lay] large MN longueur ou] *om.* BD ou] *add.* la HKN ou ... est] des climats se prent M en_1] en de D d'orient] en orient K 1016 et] et la BFHK latitude ... vers] largeur se prent en tyrant de mydi a M ou le lay] *om.* B; est D; *add.* est K ou on alant] *om.* B 1017 selonc] secun C contient] tient F 1018 qui] que F .xxv.^{e}] 25/25me DMP; .xx.^{e} HKN 1019 assavoir] ass E .viii.] 8 P; 6 D; les 8 M; les .viii. F un] .i. D 1020 est] vault M .lxxxvii.] .iiii^{xx}vii. ABFGHKN; 87 MP; 67 D milliers] ^{m} HKN demi] ½ M 1021 Alphagranus] Alphaganus A; Alfaganus BD; Alfraganus FM; Alfagranus G aucuns] *om.* B le circuite de] *om.* B 1022 car] que D ha .lvi.] a 56 DP; contient 556 M 1023 .ii.] 2 D; deux HPN .ii. tierces] ⅔ M d'un] de K d'un millier] *om.* M tant seulement et] *om.* K ainsi] aussi D; aussainsi F 1023-1024 .ii.^{c} mille et .iiii.^{c}] .cc^{m}iiii.^{c} AF; .cc.^{m} et .cccc. B; .cc.^{m} et .iiii.^{c} G; .ii.^{m} et .iiii.^{c} K; deux cens mille et .iiii.^{c} H; 200,400 D; de la terre 200,400 M; de la terre .ii.^{c} mille .iiii.^{c} P 1024 .c.^{m} et .ii.$^{c}{}_{1}$] .c.^{m} et .cc. B; .c. mil et .ii.^{c} H; 100,200 DM; cent mille et .ii.^{c} GP lieues] *add.* ou leues N .c.^{m} et .ii.$^{c}{}_{2}$] .c.^{m} et .cc. B; .c. mil et deux cens H; cent mille .ii.^{c} G; sont 100,200 M 1024-1025 et_3 ... lieues] *om.* DP

1 This line can legitimately be interpreted in two ways: (a) "longueur de lay" could refer only the latitude, that is, the length (i.e, size) of the width, or (b) "longueur et lay" could refer to both longitude and latitutde, that is, length (i.e., longitude) and width (i.e., latitude).

2 Chap. 26 (not Chap. 25), ll. 750 ff.

3 See "Source References," line 1021.

milliers – ce sont .l.^{m} et .c. lieues. Tele est la longueur du premier climat.[4]

Le secont climat est plus court; et le tiers encor plus court; et ainsi des autres, tousjours en acoursant en venant vers le septentrion, si comme il apert en figure. Mes les aucteurs ne mettent pas la certaine mesure de leur longitude.

Ceste longitude des climats fu trouvee par regarder l'eclipse de la lune, car quant la lune appert a l'eure de l'eclipse en .i. lieu plus haut et en .i. autre lieu plus bas d'un degré, il s'ensuit que le lieu ou elle appert plus haut soit plus vers orient per .i. degré du circuite de la terre.[5] Et ainsi fu il trouvé selonc plusieurs astrologiens que la distance du premier lieu habitable vers orient au desrenier vers occident est .ciiii.^{xx} degrés, c'est la moitié du circuite de la terre, et sont .cc.^{m} et .iiii.^{c} milliers, comme dit est.

1025 ce sont] qui valent M .l.^{m} et .c.] .l. mil et .c. BG; 50,100 M longueur] grandeur M 1026 Le$_{1}$ ... climat] *om.* K second] 2 D; .ii. F; deuxieme M tiers] 3 D ainsi] aussi F 1027 tousjours en acoursant] *om.* B acoursant en] acoursissant et M en$_{2}$] en la M 1028 leur] la M longitude] *add.* des autres climatz M 1029 fu] et mesmement du premier fut M regarder] regarde F regarder l'eclipse] la consideracion M lune] *add.* c'est assavoir de l'eclipse d'icelle M car] que D 1030 a l'eure] en lune K de l'eclipse] d'eclipse FGK .i.$_{1}$] *om.* BKN; un/ung ADFGHMP haut] *om.* H et ... bas] *om.* M .i.$_{2}$] *om.* BD; un/ung AFGHKP 1031 plus$_{2}$] *om.* DGK per .i.] d'un BM .i.] un/ung AFGHKNP 1032 selonc] par BM; secun C plusieurs] les K 1033 vers$_{1}$] devers K au] jusques au M est] si est M .ciiii.^{xx}] .c. et .iiii.^{xx} FG; cent quatre vins H; 180 M 1034 c'est] qui est M et$_{1}$ ... .iiii.^{c}] qui est en tout 200400 M; 200400 D .cc.^{m}] .ii.^{m} K; .ii.^{c} mille N 1034-1035 comme dit est] ut supra B

---

4 Because of the awkwardness of the roman numerals used in most manuscripts, and the variations among them, I include here a translation of lines 1023-1025 with arabic numerals (also found in mss D and M):

> "Thus the entire circumference would have 200,400 miles – that is, 100,200 leagues – and half the circumference 100,200 miles – that is, 50,100 leagues. Such is the length of the first clime."

5 This argument is faulty, since if at one place on earth the eclipse is straight overhead, it would seem lower in the sky not only from a place to the east, but also from a place to the west. Eclipses of the moon are used to determine differences in longitude, but this is generally done in terms of local time, that is, if the moon is eclipsed in one place at midnight (local time), and at a second place at one in the morning (local time), the second place is 15° (360° / 24) further east than the first.

LE .XXXIIIIE CHAPISTRE. DE LA LATITUDE DES CLIMATS

La latitude des climats est en procedant de miedi droit vers septentrion, comme dit est, et en ceste maniere. Le commencement du premier climat est par desa l'equinocial .xii. degrés et .xlv. minus, qui valent, selonc Aphagranus,[1] .vii^{c}xxii. milliers et demi – ce soit .ccclxi. leues et un quart.

Item: la latitude du premier climat est .vii. degrés et .xlv. minutes qui valent .iiii^{c}xl. milliers, petit moins. Et ainsi le met l'aucteur *De l'espere*,[2] parquoy il appert que il ne s'acorde pas a la mesure que il meismez met du circuite de la terre en son premier chapistre par estades, se il est ainsi que .viii. estades valent .i. millier, ne .vii. ne .x., ne en quelconque nombre.

Car selonc ce que il dit ou premier chapistre,[3] chascun degré ha .vii.^{c} estades qui vaudroient .lxxxvii. milliers et demi. Et ainsi la latitude du premier climat seroit environ .vi^{c}xv. milliers. Et il la met en son quart chapistre .ccccxl. milliers, aussi

1039 minus] *read* minutes[4]

1036 Le ... chapistre] *om.* HKN .xxxiiiie] 34/34me DMP climats] *add.* .xxxiiii. KN; *add.* .xxxiiii. chapitre H 1037 droit] *om.* BD 1039 .xii.] 12 DMP .xlv.] 45 DMP selonc] secun C Aphagranus] Alfraganus FM; Alfagranus DG .vii^{c}xxii.] .vii^{cc}xxii. E; 722 M; 70022 P millieurs] ^{m} K 1040 demi] ½ M soit] sont G .ccclxi.] 361 DMP; .iii^{c}lxi. B; .iii.^{c} et .li. HN; .iii^{c}li. K un] .i. G un quart] ¼ M 1041 Item] *om.* M climat] *om.* B .vii.] *marg.* C; 7 DP; sept H .xlv.] 45 DP qui valent] *om.* B 1042 .iiii^{c}xl.] .iiii^{c}xl. E; .ccccxl. AFG; 440 MP; .cccxl. B; .iiiixl. H millieurs] ^{m} K petit] ung petit M ainsi] *om.* M 1042-1043 le ... met] luy mesures M 1043 il$_{2}$] lui P meismes met] mesurent B; mesure met K de la terre] *om.* D 1044 .viii.] 8 MP; 6 D; une G .viii. ... valent] une estade valle G .i.] un/ung ABFHMNP .vii.] 7 DMP 1045 .x.] 10 DMP nombre] nienbre/menbre EC 1046 Car] que D selonc] secun C .vii.^{c}] 700 DMP; .vii. K 1047 .lxxxvii.] 87 MP; 67 D; .iiii.^{xx}vii. B demi] ½ M 1048 .vi^{c}xv.] 615 DMP milliers$_{1,2}$] ^{m} K son ... chapistre] soy D .ccccxl.] 440 DM; 540 P; .iiii^{c}xl. BHKN aussi] ainsi DM

1 See "Source References," line 1039. Al-Farghānī's gives a value of 1° = 56.66 miles.

2 Sacrobosco (*De spera*, chap. 3 [p. 139]) gives a width of the first clime as 7°45′ (12° 45′ to 20° 30′). Using al-Farghānī's value, Sacrobosco calculates the width of the first clime as 440 miles (actually 439.167). Oresme is objecting because Sacrobosco has already given a value of 700 stadia or 87.5 miles for a degree (*De spera*, chap. 1 [p. 123]), where 1 mile equals 8 stadia.

3 Oresme is referring to Sacrobosco, chapter 1, and four lines later also to Sacrobosco, chapter 4 (actually 3).

4 Here in this chapter and in the following chapters, base ms E writes "minus" or "minues" for "minutes"

comme fait Alphagranus.[5] Et je cuide que ce soit la meilleur mesure. Et selonc ceste mesure, chascun degré du circuite de la terre ha ta[nt] seulement .lvi. milliers et .ii. tiers, comme dit est.[6] Et selonc ceste mesure, le circuite de la terre n'est pas si grant comme dit l'autteur *De l'espere* et comme il fu dit ou .xxve chapistre.[7] Je di donques le premier climat a .ccxx. lieuez de lay.[8]

f. 20^{v} Le secont ha .vii. degrés qui valent .cccc. milliers – ce sont / .ii.^{c} leues.[9]

Le tiers, .vi. degrés et .x. minues qui valent .cccl. milliers – ce sont .clxxv. leues.

Le quart, .v. degrés et .xx. minues qui valent .ccc. milliers – ce sont .cl. leues.

Le quint est lay .iiii. degrés et demi qui valent .cclv. milliers – ce sont .cxxvii. leues et demie.

Le sixte, .iii. degrés et .xlv. minues qui valent .ccx. milliers – ce sont .cv.

1049 Alphagranus] Alfraganus FM; Alfagranus DG soit] seroit M mesure] *om.* B selonc] secun C 1050 ha] *om.* B tant] ta E .lvi.] 56 DMP .ii.] 2 D; deux BHP 1050-1051 .ii. tiers] ⅔ M 1051 selonc] secun C 1052 .xxve] 25/25me DMP 1053 donques] *add.* que BDGMP .ccxx.] 220 MP; 420 D; .ii^{c}xx. HKN lay] large MN 1054 second] 2 D; .ii. F; deuxieme M .vii.] 7 DMP; sept H .cccc.] 400 MP; .iiii.^{c} BFKN; quatre cens H .ii.^{c}] 200 MP; deux cens H 1054-1055 .cccc. ... valent] *om.* D 1055 tiers] *add.* a B .vi.] 6 MP; six H .x.] 10 MP; dix H .cccl.] 350 DMP; iii^{c}l. BFK; trois cens .l. H; .iii.^{c} N milliers] ^{m} K .clxxv.] 175 BDMP; cent .lxxv. G; cent soixante quinze H 1056 leues] *om.* B 1057 quart] 4 D; *add.* climat si a M .v.] 5 DMP; a cinq B .xx.] 20 DMP; vint H .ccc.] 300 DMP; .iii.^{c} A; trois cens HN milliers] ^{m} K .cl.] 150 DBMP; cens .l. H 1058 quint] 5 D; cinquieme M est lay] ha B; a de large M .iiii.] 4 DMP demi] ½ M .cclv.] 255 BDMP; .ii^{c}lv. FN; deux cens .lv. H .cxxvii.] 127 BMP; 717 D; cent .xxvii. H; .clv. K 1059 demi] ½ M 1060 sixte] 6/6me DP; .vi./.vi.^{c} BK; sixieme a M .iii.] 3 DMP; trois H; .iiii. G degres] *add.* de large M .xlv.] 45 DP; 48 M; .xlviii. F minues] mi|minus A .ccx.] 210 BDMP; .cx. F; deux cens .x. H; .ii^{c}x. KN milliers] ^{m} K .cv.] 105 BMP; 150 D; cent .v. H; .clv. N

5 See "Source References," line 1049. Using Sacrobosco's value of 87.5 miles per degrees (from his chap. 1), the width of the first clime would be 678.125 miles (not 615 as in the text); but, as Oresme points out, Sacrobosco adopts al-Farghānī's 440 mile width in his later chapter – actually chap. 3 (p. 139) in the modern edition rather than "chap. 4." (The division into chapters would depend on which ms Oresme was consulting.)

6 Above, al-Farghānī, line 1039.

7 As already noted, there is an inconsistency in Sacrobosco and Oresme is stating that the former's earlier value for the circumference of the earth is wrong. The reference to "Chap. 25" should be to Oresme's Chap. 26.

8 These values and measures are summarized in the appendix on "Climes", section 4.

9 The given values in miles, here and following, are within 1% of their calculated values.

lieues. Le septime, .iii. degrés et .xv. minues qui valent .clxxxv. milliers – ce sont .xcii. leues et demie.

Et ainsi [la] latitude de tous les climas ensemble est .xxxviii. degrés et .xlv. minues[10] qui valent .ii^{m}cxl. milliers – ce sont .mlxx. lieues. Ceste mesure de la latitude des climas fu trouvee par considerer en divers lieus l'elevacion du pole ou d'une estoille en la maniere que fu trouvé le circuite de la terre, si comme il appert ou .xxvie chapistre.[11]

1061 septime] 7/7me DP; .vii./.viie BFHK .iii.] 3 DP; a 3 M .xv.] 15 DMP; quinze B .clxxxv.] 185 BDMP; .lxxxv. F; cent .lxxxv. H; .clx. et .xxv. N 1062 .xcii.] 92 DMP; 102 B; cent .ii. H; .x. cens .ii. G; .cii. AKN et demi] ½ M 1063 Et] *add.* qui D la] *om.* EC est] *om.* B; *interlin.* C .xxxviii.] 38 BMP; 36 *corr. to* 38 D .xlv.] 45 BDMP 1064 qui] *om.* H .ii^{m}cxl.] 2140 BMP; 2100 D; deux mil cent .xl. H .mlxx.] 1070 DMP; mil/mile .lxx. FGHKN; mil et septente B lieues] eaues G mesure] maniere M la] *om.* ADFM 1065 en ... lieus] *om.* M l'elevacion] la levacion G 1066 que] qui D le circuite] la circonference M il] *om.* F appert] apper H 1067 .xxvi.^{e}] 26/26me DMP chapistre] *add.* Cy enseust la figure B

10 Most mss agree on 38° 45′ for the width of all the climes, but in fact the width (adding the widths for the 7 climes together) should be 37° 45′.

11 Chap. 26, ll. 746-764.

DIAGRAM: CHAPTER 34

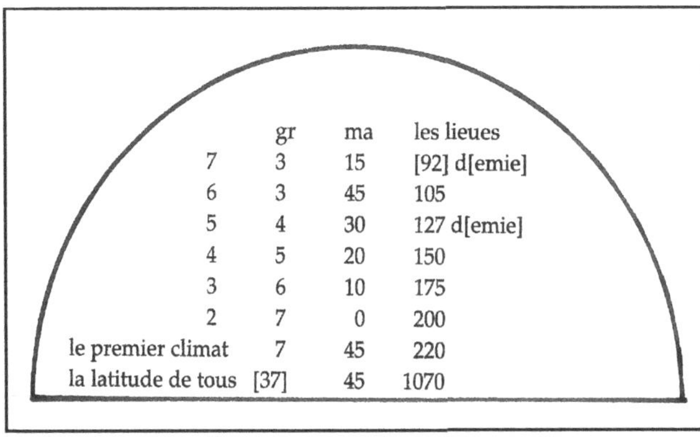

La figure de la latitude des climes

Complete diagram: E CP
No diagram (or space only): ABDFGHKMN
Chart: *see below*

92] 912 EC demie$_{1, 2}$] d ECP 37] 38 ECP

CHART: CHAPTER 34

La latitude des climas

	dégres	minus	milliers	lieues
du premier	.vii.	.xlv.	.ccccxl.	.ccxx.
du second	.vii.	o	.cccc.	.cc.
du tiers	.vi.	.x.	.cccl.	.clxxv.
du quart	.v.	.xx.	.ccc.	.cl.
du quint	.iiii.	.xxx.	.cclv.	.cxxvii. et demie
du sixte	.iii.	.xlv.	.ccx.	.cv.
du vii^{e}	.iii.	.xv.	.c iiii^{xx} v.	.iiii^{xx} xii. et demie
de tous ensemble	.xxxvii.[12]	.xlv.	.ii^{m} cxl.	.mlxx.

[*Complete chart*:] ABDFGHKMN

1 *pre* La] *add.* Cy ensieust la figure B 1 La ... climas] La latitude des climas | latitude H 2 degres] deg's F minus] mīn F; minutz B 3 du premier] premier K; le premier D; .i. H .vii.] 7 DM .xlv.] 45 DM .ccccxl.] .cccxl. ABKF; .iii^{c} x.l G; .iiii^{c} xl. H; 320 D; 440 M .ccxx.] .ii^{c} xx. GH; .ccc. K; 220 DM 4 du second] secund K; .ii.^{c} H; le 2 D vii] 7 DM cccc] ccc N; iiii^{c} GH; 400 M; 300 D cc] ii^{c} GH; 200 DM 5 du tiers] tiers K; le 3 D; .iii. H .vi.] 6 DM .x.] 10 DM .cccl.]. iii.^{c} l GH; 350 DM .clxxv.] 175 DM 6 du quart] quart K; le 4 D; .iiii. H . v.] 5 DM . xx.] 20 DM .ccc.; .iii^{c} GH; 300 DM .cl.] 150 DM 7 du quint] quint K; du .v.^{e} G; v. H; le 5 D .iiii.] 4 DM .xxx.] 30 DM .cclv.] .ii^{c} lv. G; .ii^{c} xlv. H; 255 D; 245 M . cxxvii.] .cxxxii. F; .cxviii. H; 127 DM demie] d DF; ½ K 8 du sixte] du .vi.^{e} G; .v.i H; .vi.^{e} K; le 6 D .iii.] 3 DM .xlv.] 45 DM .ccx.] .ii^{c} x. GH; 210 DM .cv.] 150 D; 105 M 9 du .vii.^{e}] .vii. H; .vii.^{e} K; du sept^{e} M; du septiesme ABN; le 7 D .iii.] 3 DM .xv.] 15 DM .c iiii^{xx} v.] .cxiii. B; 185 DM .iiii^{xx} xii.] 92 DM demie] d DF; ½ M 10 de tous ensemble] tous ensemble K .xxxvii.] .xxxviii. ABFGHKN; 37 M; 38 D .xlv.] 45 DM . ii^{m} cxl.]. ii.^{m} cent .xl. N; 2140 DM .mlxx.] mil .lxx. BN; 1070 DM

12 All manuscripts (except M) have .xxxviii. / 38, but the actual amount should be .xxxvii. / 37.

LE .XXXV. CHAPISTRE. DE LA QUANTITÉ DU PLUS LONG JOUR EN CHASCUN CLIMAT

Si comme il apparut par le .xviiie chapistre, tant plus s'esloigne l'en de l'equinocial en venant vers septentrion, tant sont les jours plus longe en esté et [plus] cours en yver[1] si que au commencement du premier climat, le plus long jour naturel est de .xii. heures et .xlv. minus d'une heure.[2] Et en la fin du premier climat – c'est au commencement du secont – le plus longe jour est de .xiii. heures et de .xv. minus. Ou commencement du tiers le plus longe jour est de .xiii. heures et .xlv. minus; ou f. 21^{r} commencement du quart, de .xiiii. heures et de .xv. minus; ou commen/cement du quint, de .xiiii. heures et de .xlv. minus; ou commencement du sixte, de .xv. heures et de .xv. minus; ou commencement du septisme, de .xv. heures et de .xlv. minus; en la fin du septisme, de .xvi. heures et de .xv. minus. Et par ceci appert que en chascun climat, du commencement siques en la fin, quant au plus long jour, a difference de demie heure. Et du commencement du premier siques en la fin du septisme ha difference de .iii. heures et demie. Et ou regart de ceste cause et des .vii. planetes fu la terre bien habitable ainsi divisee en .vii. climas.

1073-1079 minus] *read* minutes

1068 Le ... chapistre] *om.* HKN .xxxv.] 35/35me MP quantite] qualite N 1068-1105 Le ... jour] *om.* D 1069] climat] *add.* .xxxv. KN; *add.* .xxxv. chap. H 1070 apparut] appert K .xviii.^{e}] 18me MP plus] *om.* K l'en] l'an G 1071venant] alant K plus$_2$] *om.* ECP 1072 si que] *om.* P; sic que B; si comme FM naturel] *om.* GK 1073 .xii.] 12 MP .xlv.] 45 MP c'est] est N 1073-1074 .xii. ... de$_1$] *om.* F d'une ... secont] *om.* M 1074 jour] *add.* du second M est] *om.* F .xiii.] 13 MP de$_2$] *om.* GM .xv.] *illeg.* B; 15 MP 1075 Ou] En K .xiii.] 13 MP et] *add.* de HK .xlv.] 45 MP 1076 quart] *add.* le plus long jour est M .xiiii.] 14 MP de$_2$] *om.* BFGKM .xv.] 15 MP ou commencement] *om.* K 1077 .xiiii.] 14 MP de$_2$] *om.* BFG .xlv.] 45 MP; .xv. F sixte] 6me P; .vi.^{e} B; sexieme M de$_3$] *om.* B .xv.] 15 MP 1078 de$_1$] *om.* ABGKM .xv.$_1$] 15 MP .xv.$_1$ ... .xlv.] .~~xv~~. .xlviii. HN ou commencement] *om.* K septisme] 7me P; .vii./.vii.^{e} BF .xv.$_2$] 15 MP; .xvii. K .xlv.] 45 MP 1079 en ... minus] *om.* M septisme] .vii.^{e} K .xvi.] 16 P de$_2$] *om.* ABFG .xv.] 15 P que] *om.* P 1080 siques] *om.* P; jusques KMN en] a BM long jour] *om.* M 1081 siques] jusques BHKMNP en] a BM 1082 septisme] .vii.^{e} AGK; 7^{e} M .iii.] 3 P; trois BHM .vii.] 7 P; sept BM 1083 ainsi] *om.* BM .vii.] 7 MP

1 Chap. 18, ll. 493-495.

2 Data summarized in the appendix on "Climes", section 4.

Et pour ce, est le premier climat plus lay que le secont, et le secont plus lay que le tiers et ainsi ensuiant, si comme il appert par le chapistre precedent.[3] Car pour trouver le plus grant jour plus longe de demie heure, il couvient passer plus grant distance en venant vers septentrion quant l'en est pres de l'equinocial, et ou premier climat que ou secont climat, et ou secont que ou tiers, et ainsi aprés. Et ce est pour la roondesce de la terre et pour obliquité de l'ozizon et pour obliquité du zodiaque.

Item: l'aucteur *De l'espere*[4] et aucuns autres mettent que le plus long jour ou milieu du premier climat est de .xiii. heures; et ou milieu de second, de .xiii. heures et demie; du tiers, .xiiii. heures; du quart, .xiiii. heures et demie; du quint, .xv. heures; du sixte, .xv. heures et demie; du septisme, .xvi. heures. Mes ce n'est pas bien assigné le milieu, car comme il est touchié devant.

Pour trouver le plus grant jour plus long [de] demie heure, ou du quart d'une heure, il couvient plus passer de distance en venant vers septentrion, quant l'en est pres de l'equinocial que quant l'en est plus loing. Et pour ce, a trouver plus long jour du quart d'une heure au commencement en la premiere partie du premier climat, il couvient passer plus de terre que en la fin d'icelui. Par quoy il s'ensuit que le lieu ou le plus long jour est de .xiii. heures est plus pres de la fin du premier climat que du commencement, et ainsi n'est ce pas le milieu. Et semblablement est il des autres climas.[5]

1084 lay$_1$] large MNP plus lay$_2$] *om.* M; plus large NP 1085 ainsi] *add.* en P si] *om.* KN il] *om.* F 1086 grant$_2$] de G 1087 l'en] on P 1088 climat$_2$] *om.* BFGKM apres] desautres B; ensuyant K 1089 la$_1$] les M obliquite$_{1,2}$] l'obliquite ABCKM et$_2$ ... zodiaque] *om.* N du] aussi du K 1091 .xiii.$_1$] 13 MP second] *add.* est M .xiii.$_2$] 13 MP 1092 demie$_1$] ½ M du$_1$ ... heures$_2$] *om.* G tiers] *add.* de ABFKM .xiiii.$_1$] 14 MP quart] *add.* de BGKM .xiiii.$_2$] 14 P heures$_2$] *om.* AF et demie$_2$] ½ M quint] *add.* de BKM .xv.] 15 MP 1093 sixte] 6me P; .vi./.vi.^{e} de BHKN; *add.* de KM .xv.] 15 MP; quinze B heures$_2$] *om.* AFG demie] ½ M septisme] .vii.^{e} HKN; *add.* de BKM .xvi.] 16 P; 16 et ½ M 1095 long] grant B de] *om.* ECP 1096 passer] *add.* plus H l'en] l'an G 1097 que] ne N l'en] l'an G; l'en en A plus$_1$] *om.* K trouver] *add.* le K 1097-1098 plus$_2$ ... du$_1$] d'une B; ung jour plus long d'ung M 1098 d'une] de BM en] ou a B; ou en AFGHKMN 1099 Par] Pour BP 1100 .xiii.] 13 MP est$_2$] et AGH; et est N 1101 est il des] des dix HKN

3 Chap. 34, ll. 1052-1062.

4 See "Source References," line 1090.

5 Just as with the climes themselves where the lines which define the regular sequence of the times of the longest days (half-hour intervals) are closer and closer the further north one goes, so within a clime the division of time, say into quarter hours, will be such that the first quarter is wider than the second quarter (or with 10-minute intervals, the first interval will be wider than the second) and the second than the third and so on. (See also the appendix on "Climes", section 5.)

Doncques ne sont pas les milieuz des climatz bien assignéz par l'aucteur *De l'espere*. Et convendroit user des subtilités de geometrie qui sont en Ptholomee qui les voudroit assigner a point selonc la quantité du plus long jour.[6]

1103 milieuz] lieux G bien] *marg.* C l'aucteur] l'auctorite K 1104 geometrie ... Ptholomee] theorique K; theologie N; astrologie et theologie H qui$_1$ ... Ptholomee] *om.* FGHN Ptholomee] Ptolomee C 1105 selonc] secun C jour] *add.* Cy enseust la figure de la quantite du plus long jour en chacun climath ici B

f. 21v / DIAGRAM: CHAPTER 35

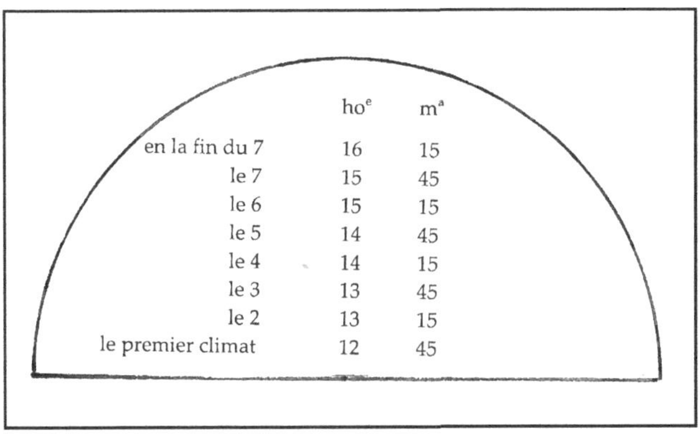

La figure de la quantité de plus lonc jour
en chascun climat au commencement

Complete diagram: E CP
No diagram (or space only): D
Alternative chart: BFGHKMN

ho^{e}] hoe C le 2] le 2me P

6 In lines 1103-1105 Oresme is saying that Sacrobosco is wrong to treat divisions (by time-periods – 30 minutes, 15 minutes, etc.) as equal widths within a clime, but that one must use the geometry followed by Ptolemy in the *Almagest* to find the right widths (in degrees) of these time units. (See the appendix on "Climes", section 5.)

CHART: CHAPTER 35

La quantité du plus long jour en chacun climat

Au commencement		
du premier	heures .xii.	minus .xlv.
du second	.xiii.	.xv.
du tiers	.xiii.	.xlv.
du quart	.xiv.	.xv.
du quint	.xiv.	.xlv.
du .vi.^{e}	.xv.	.xv.
du .vii.^{e}	.xv.	.xlv.
En la fin du .vii.^{e}	.xvi.	.xv.

Complete chart: BFGHKMN

2 Au] *om.* K 2-3 Au ... premier] .i. commencement .i. H 3 du premier] *add.* climat G .xii. 12 M .xlv. 45 M 4 du second] Au commencement du second GM; .ii. commencement .ii. H .xiii. 13 M .xv. 15 M 5 du tiers] Au commencement du tiers M; .iii. commencement .iii. H .xiii. 13 M .xlv. 45 M 6 du quart] Au commencement du quart M; .iiii. commencement .iiii. H .xiiii. 14 M .xv. 15 M 7 du quint] Au commencement du quint M; .v. commencement .v. H .xiiii. 14 M .xlv. 45 M 8 du .vi.^{e}] du sixte FN; Au commencement du sex^{e} M; .vi. commencement .vi. H .xv. 15 M .xv. 15 M 9 du .viie] du septieme N; Au commencement du sept^{e} M; .vii. commencement .vi. H .xv. 15 M .xlv. 45 M 10 En la] A la M .viie] septiesme GHMN .xvi. 16 M .xv. 15 M

LE .XXXVIE CHAPISTRE. DE L'ELEVACION DU POLE EN CHASCUN CLIMAT

Le pole n'est point eslevé a ceulz qui sont sous l'equinocial, mes il est eslevé partout aillours, et est plus eslevé sus l'ozizon de tant comme l'en s'eslonigne plus de l'equinocial et que l'en ha l'ozizon plus oblique, si comme il fu dit ou .xve chapistre.[1]

Or appert il par le .xvie chapistre [que] aussi grande est l'elevacion du pole sus chascun ozizon comme est la distance d'entre cenith et l'equinocial.[2] [Et il appert par le .xxxiiie chapistre que la cenith] ou commencement du premier climat est eslevé par .xii. degrés et .xlv. minus,[3] par quoy il s'ensute que le pole est eslevé au commencement du premier climat .xii. degrés et .xlv. minus.

Item: ajouxtés avec ce la latitude du premier climat, laquelle est .vii. degrés et .xlv. minus, vous aures la distance de l'equinocial et le cenith de la fin du premier climat. Et par consequent, l'elevacion du pole en la fin du premier climat et ou commencement du secont est .xx. degrés et demi.

Item: ajoustés aveques tout ceci la latitude du secont climat, et vous aures l'elevacion du pole en la fin du secont climat et ou commencement du tiers et est .xxvii. degrés et demi.

Item: par ceste voie trouveres l'elevacion du pole au commencement du quart climat .xxxiii. degrés et .xl. minus.

1114-1138 minus] *read* minutes

1106 Le ... chapistre] *om.* HKN .xxxvie] 36/36me DMP 1107 climat] *add.* .xxxvi. KN; *add.* .xxxvi. chap. H 1108 point] *om.* D 1109 et] mais il B l'en] l'an G l'en s'eslongne] on eslongne D plus] *om.* M 1110 l'en] l'an G l'ozizon] orizon DM .xv.^{e}] 15/15eme DMP 1111 Or] Que aussi D .xvi.^{e}] 16/16eme DMP que] car ECP l'elevacion] la levacion G 1112 est] *om.* KN d'entre] *add.* le M et l'equinocial] *interlin.* C 1112-1113 Et$_2$ ... cenith] *om.* CE 1113 .xxxiii.^{e}] 33/33^{e} DM; .xxiii.^{e} KN climat] *om.* D 1113-1114 esleve par] loing de l'equinocial M 1114 .xii.] 12 DMP et] *om.* M .xlv.] 45 DMP 1114-1115 par ... climat] de B par ... minus] *om.* M 1115 .xii.] 12 DP .xlv.] 45 DP 1116 ce la] celle N .vii.] 7 DMP 1117 .xlv.] 45 DMP de$_1$] d'entre FHKMN 1118 Et$_1$ ... et$_2$] *om.* M l'elevacion] la levacion G du$_2$ ... climat] *om.* D ou] *om.* K 1119 secont] 2 D; *add.* et M .xx.] 20 DM; .xxx. CE; 30 *corr. to* 20 P et demi] ½ M 1120 avecques] ouvec G tout] *om.* D secont] 2 D 1121 l'elevacion] la levacion EG secont] 2 D climat] *om.* B tiers] 3 D et$_2$] *om.* HN 1122 .xxvii.] 27 DM et demi ½ M 1123 l'elevacion] la levacion G quart] 4 D 1124 climat] *om.* K .xxxiii.] 33 P; 32 D; qui est 33 M .xl.] 40 DM

1 Chap. 15, ll. 395-403.

2 Chap. 16, ll. 411-416.

3 Chap. 34 (not 33), ll. 1038-1039.

f. 22r Item: au commencement du quint, / .xxxix. degrés.

Item: au commencement du sixte, .xliii. degrés et demi.

Item: au commencement du septisime, .xlvii. degrés et .xv. minus; et en la fin du septisme, .l. degrés et demi.

Item: adjoustés a l'elevacion du pole au commencement du premier climat la moitié de la latitude du premier climat et vous aurez l'elevacion du pole ou milieu du premier climat. Et par semblable voie aures vous l'elevacion du pole ou milieu de chascun climat.

Et trouveres que l'elevacion du pole ou milieu du premier climat est .xvi. degrés, .xxxvii. minus et demi.

Item: ou milieu du second climat .xxiiii. degrés; ou milieu de tiers, .xxx. degrés et .xxxv. minus; et ou milieu du quart, .xxxvi. degrés et .xx. minus; ou milieu du quint, .xli. degrés .xv. minus; ou milieu du sixte, .xlv. degrés .xxii. minus et demi; et ou milieu du septisme .xlviii., degrés .lii. minus et demi. Et ainsi avon nous l'elevacion du pole en chascun climat au commencement et au milieu [et] en la fin.

1125 quint] 5 D .xxxix.] 39 MP; 29 D 1126 Item] *om.* F sexte] .vi.e B; sixiesme M .xliii.] 43 MP et demi] ½ M 1126-1127 Item ... minus] *om.* D 1127 septisime] .vii.e BHK .xlvii.] 47 MP; .xliii. K .xv.] 15 MP et en] en HN; et a M; item en K 1128 septisme] 7 D; .vii.e HK; .viii.e N .l.] 50 DMP demi] ½ M 1029 l'elevacion] la elevacion G au commencement] *om.* K; du premier climat et vous auvez l'elevation du pole au milieu HN (*copying part of the following lines*) climat] *add.* avecques M 1130 l'elevacion] la levacion G pole] *add.* au premier climat et B 1130-1131 et ... climat] *om.* P 1131 l'elevacion] la levacion G 1132-1133 de ... milieu] *om.* D 1133 l'elevacion] la levacion G est] *rep.* N .xvi.] 16 DMP; .xvii. A; .xxvi. N 1134 .xxxvii.] 37 DMP dem i] ½ M 1135 second] 2 D; chascun HN clim at] *add.* l'elevacion du pole si est M .xxiiii.] 24 DMP degres] Item D tiers] 3 D .xxx.] 30 DMP 1136 .xxxv.] 35 BDMP quart] 4 D .xxxvi.] 36 BDMP .xx.] 20 DMP 1137 quint] 5 D .xli.] 41 BDMP .xv.] 15 BDMP sixte] 6/6me DP; .vi.e B; sexiesme M .xlv.] 45 BDMP .xxii.] 22 BDMP demi] ½ M 1137-1138 et$_2$ ... demi] *om.* D 1138 septisme] 7me P; .vii.e HKN .xlviii.] 48 BMP .lii.] 52 BM; 22 P demi] ½ M nous] *om.* CFGHKMN 1139 l'elevacion] *om.* D; la levacion G au commencement et] *om.* H et$_2$] *om.* ECP; ou B en] aussi a la M

DIAGRAM: CHAPTER 36

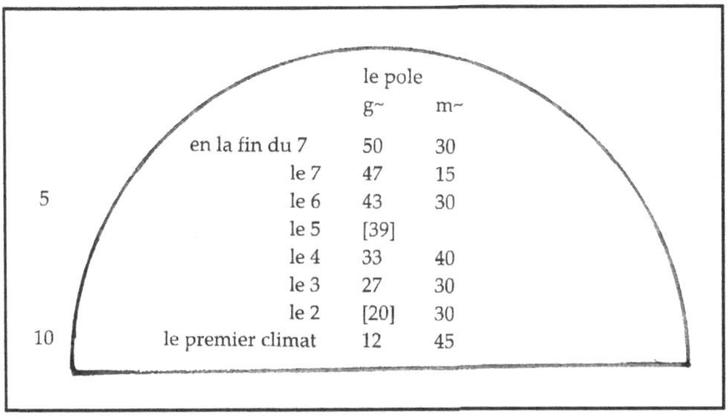

La figure de la levation du pole en chascun climat

Complete diagram: E CP
Chart: see below

4 47] 45 P 6 39] 30 E 9 20] 30 ECP

CHART: CHAPTER 36

L'elevacion du pole en chascun climat

	Au commencement:		Ou milieu:			En la fin:	
	degrés	minus	degrés	minus	secons	degrés	minus
du premier climat	.xii.	.xlv.	.xvi.	.xxxvii.	.xxx.	.xx.	.xxx.
du second	.xx.	.xxx.	.xxiiii.	.0.	.0.	.xxvii.	.xxx.
du tiers	.xxvii.	.xxx.	.xxx.	.xxxv.	.0.	.xxxiii.	. xl.
du .iiii.^{e}	.xxxiii.	.xl.	.xxxvi.	.xx.	.0.	.xxxix.	.0.
du .v.^{e}	.xxxix.	.0.	.xli.	.xv.	.0.	.xliii.	.xxx.
du .vi.^{e}	.xliii.	.xxx.	.xlv.	.xxii.	.0.	.xlvii.	.xv.
du .vii.^{e}	.xlvii.	.xv.	.xlviii.	.lii.	.xxx.	.l.	.xxx.

Complete chart: BDFGHKMN

2 Au] *om.* D; en H; ou GN   Ou] *om.* D   En la] *om.* D; a la M   En la fin] *om.* K; de fine B   3 degres$_{1,2,3}$] gradus B; gres M; g~ D   minus$_{1,2,3}$] m M; m~ D   secons] *om.* M; 2~ B   4-5 du ... climat] .i. F   5 climat] *om.* HKMN   xii.] 12 BDM   .xlv.] 45 BDM   .xvi.] 16 BDM   .xxxvii.] *corr. from* .xxxviii. K; 37 BDM   .xxx.$_{1}$] *om.* M; 30 D   .xx.] 20 BDM   .xxx.$_{2}$] xx K; 30 BDM   6 du second] .ii. F; du 2 D   .xx.] 20 BDM   .xxx.$_{1,2}$] 30 BDM   .xxiiii.] 24 BDM   .0.$_{2}$] *om.* M   .xxvii.] 27 BDM   7 du tiers] .iii. F; du 3 D   .xxvii.] 27 BDM   .xxx.$_{1,2}$] 30 BDM   .xxxv.] 35 BDM   .0.] *om.* M   .xxxiii.] 33 BDM   . xl.] 40 BDM   8 du .iiii.^{e}] .iiii. F; du 4 D; du quart BHKN   .xxxiii.] 33 BDM   .xl.] 40 BDM   .xxxvi.] 36 BDM   .xx.] 20 BDM   .0.$_{1}$] *om.* M   .xxxix.] 39 BDM   .0.$_{2}$] 10 D   9 du .v.^{e}] .v. F; du 5 D; du quint BHKN   .xxxix.] 39 BDM   .0.$_{1}$] 10 D   .xli.] 41 BDM   .xv.] .xx. F; 15 BM; 25 D   .0.$_{2}$] *om.* M   .xliii.] 43 BDM   .xxx.] 30 BM; 40 D   10 du .vi.^{e}] .vi. F; du 6 D; du sexte HN; du siste K   .xliii.] 43 BM; 42 D   .xxx.] 30 BM; 40 D   .xlv.] 45 BDM   .xxii.] 22 BDM   .0.] *om.* M   .xlvii.] .xlviii. K; 47 DM; 42 B   .xv.] 15 BDM   11 du .viie] .vii. F; du 7 D; du septieme HKN   .xlvii.] 47 BDM   .xv.] 15 BDM   .xlviii.] 48 BDM   .lii.] 52 BDM   .xxx.$_{1}$] *om* M; 30 D   .l.] 50 BDM   .xxx.$_{2}$] 30 BDM

LE .XXXVIIE CHAPISTRE. DE LA QUANTITÉ DE TERRE HABITABLE

Selonc les astrologiens, la longueur ou longitude de terre habitable est la moitié du circuite de la terre, si comme il fu dit ou .xxxiii. chapistre.[1] Et la latitude est le lay f. 22^{v} de tous les / climats ensemble laquelle est .xxxviii. degrés et .xlv. minus. Et, selonc Albategni,[2] encores a il habitacion oultre les climats vers midi siques a l'equinocial et aussi vers septentrion en tant que la latitude de terre habitable est de dessous l'equinocial siques a .lx. degrés vers septentrion – ce est la sixte partie du circuite de la terre. Et par ceci vuelt conclurre Albategni[3] que tant seulement la .xiie partie de terre est habitable, car qui prent la moitié d'une chose du long et la sixte partie du lé, telle porcion est la douzisme partie de la chose, pour ce que la moitié de la sixte partie, c'est la douzisme partie; aussi comme qui prendroit d'un champ la moitié en long et la sixte partie en lai, ce seroit la douzisme partie.

Mais, sauve sa reverence, combien que ceste regle tiengne et ait lieu en chose plaine et droite, toutevoies elle ne conclude pas a son propos en superfice de chose roonde; et partie de tele superfice est la terre habitable.

1143 minus] *read* minutes

1140 Le ... chapistre] *om.* HKN .xxxvii.^{e}] 37/37me MP; 27^{e} D De ... habitable] De la terre habitable et sa quantite D de$_{2}$] de la G habitable] *add.* .xxxvii. KN; *add.* .xxxvii. ca. H 1141 Selonc] Secun C de] de la D 1142 .xxxiii.] 33 M; 23^{e} D; .xxx. et .iii. N lay] *add.* ou largeur N 1143 .xxxviii.] 38 DMP; .xxxvi. B .xlv.] 45 DMP selonc] secun C 1144 Albategni] Abbategni F; Albathegny G siques] jusques HKMNP 1145 aussi] *om.* ADFGHKMN; *interlin.* C que] comme K de$_{2}$] *om.* BGKN; *interlin.* C de dessous] du dessoubz de M 1146 siques] jusques HKMNP .lx.] 60 P; 6 D; quarante M ce] et M sixte] 6/6me DP; sexiesme M; tierce G 1147 la$_{1}$] *om.* N Albategni] Abbategni FK; Abbayteg' D; Albathegny G .xiie] 12/12me DMP 1148 de] de la GN car] que D chose] *add.* tout B sixte] 6/6me DP; sixiesme M partie] *om.* B 1149 lé] large M douzisme] 12/12me DMP; .xiie BGHN 1149-1150 telle ... partie$_{1}$] *om.* K 1149-1151 de$_{1}$ ... partie] *om.* M 1150 sixte] 6 D; .vi.^{e} B c'est] est DN; si est G douzisme] 12/12me DP; .xiie BGHKN aussi] ainsi FM prendroit] *add.* d'une chose ou B d'un] 1 D 1150-1151 la moitié] *ms* L *begins* 1151 en$_{1}$] du GLM sixte] 6/6me DP; .vi.^{e} B; sixiesme M partie$_{1}$] *om.* B en$_{2}$] du GL; selon le large M douzisme] 12/12me DMP; .xiie BGHKLN 1152 tiengne ... lieu] ait lieu de tenir B 1153 conclude] conclut M a] en F 1153-1154 de ... superfice] *om.* D; *rep.* N 1154 et ... habitable] *om.* GL; dont la terre habitable est partie de telle superficie M la] *om.* B

1 Chap. 33, ll. 1032-1035,

2 Al-Battānī, in his geographical description of the world (chap. vi), mentions people living in Sri Lanka, well south of the Tropic of Cancer. See "Source References," line 1144.

3 This calculation is in the same chapter (vi) of al-Battānī's work.

Item: par le consideracion du .xxxi. chapistre, la terre habitable est une des quartes parties de la terre.[4] Et ainsi le met Albategni.[5] Et qui prent tout le long de ceste quarte partie et .lx. degrés de lé, a commencier devers l'equinocial, si comme prent Albategni, il appert clerement que c'est plus de la moitié du quart de la superfice de la terre. Par quoy il s'ensuit que plus assés de l'otizisme partie de la terre est habitable.

Item: Albetagni[6] met habitables les .ii. pars ou les .ii. tiers de la latitude de la quarte partie de la terre, et encor de la part la ou celle quarte partie est la plus large. Par quoy il s'ensuit de necessité que plus des .ii. pars ou des .ii. tiers de celle quarte partie soit habitables. Et les .ii. tiers du quart sont une sixte, et ainsi est habitable plus que la sixte partie de la terre, et par consequent plus que .ii. douziesmes. Mes la certaine mesure n'en puet pas estre proprement sceue, si comme il fu dit de la longitude des climas ou .xxxiii. chapistre.[7]

1159 otizisme] *read* octisme

1155 .xxxi.] 31/31me MP; 3e D; .xxie BFHKN est] *add.* en BGHKLMN 1156 quartes] .iiii. HN Albategni] Abbategni F; Albathegni GLM 1157 et] est D .lx.] 60 DMP lé] large M si] *om.* HKN 1158 Albategni] Abbategni F; Albathegny GL il] qu'il N appert] pert G 1159 asses de l'octisme] de la moitié 8e M l'otizisme] l'onzisme EDP; la 6 D; la .viiie BHKN la$_{2}$] *om.* K 1160-1162 est ... terre] *om.* M 1161 Albetagni] Albategni ABFKNP; Albathegni GL habitables] *add.* .ii. lays ou B .ii.$_{1}$] 2 D; deux HNP .ii.$_{2}$] *om.* D; deux BHP tiers] 3 D la$_{1}$] *om.* N 1162 quarte$_{1}$] 4 D la$_{3}$] *om.* D ou] ou est L quarte$_{2}$] 4 D est] est de D 1163 necessite] *lacuna* D des$_{1}$] *om.* P; de N .ii.$_{1}$] deux DHNP .ii.$_{1}$ ... tiers] ⅔ M des$_{2}$] de N .ii.$_{2}$] *om.* D; deux BHNP 1164 soit] soient FGHKLN; sont M habitables] *add.* Et le ⅓ de celle quarte partie est inhabitable M .ii.] 2 D; deux BHNP .ii. ... quart] ⅔ de ¼ M du] d'ung P une sixte] 1/6 MP; un .vi.^{e} B 1164-1165 et ainsi ... partie] *om.* M 1165 sixte] 6/6me DP; .vi.^{e} B que$_{2}$] *om.* HL .ii.] de 2 D .ii. douziesme] 2/12 P; 1/12 M; .ii. .xii.^{es} BGKL; deux .xii.^{es} HN 1166 la$_{1}$ ... mesure] ce est lar'toyne(?) B n'en] ne ABFGLM 1167 .xxxiii.] 33me P; 32^{e} M; .xxxii.^{e} BFGL; 22 D; .xxiie HKN

4 Chap. 31, ll. 946-949.

5 The source for al-Battānī is the same (chap. vi) as noted for line 1144.

6 Again, see the Source Reference for line 1144.

7 Chap. 33, ll. 1026-1029.

LE .XXXVIII.[E] CHAPISTRE. DES HABITACIONS QUI SONT DEHORS LES CLIMAS

Les climas pris ensemble sont terminés par .iiii. lignes, si comme il appert en la figure ou .xxxi.[e] chapistre,[1] desquelles lignes une commence le premier climat [et] est par desa l'equinocial .xii. degrés et demi. Oultre ceste ligne mettent les aucteurs plusieurs habitacions et par espeial la desreniere habitacion congneue par de la est
f. 23r une isle appellee Taprobane,[2] et est a l'opposite de Ynde vers orient, selonc / Albategni.[3] Et selonc les hystoriens, elle est oultre et par de la l'equinocial, car les habitans de celle isle, quant ils regardent vers leur midi, il ont soleil levant en orient a destre et occident a senestre, et nous a contraire. Et si ne povent veoir l'estoille du nort ne celles qui en sont pres, et voient les estoilles qui nous sont perpetuelment mucíees entre lesquelles en ha une claire qui est appeles "Canopos",[4] selonc Plinius[5] et Solinus,[6] et est grande et blanche, non pas trop ardant si comme dit Albumazar[7] qui en fait mencion et appele cil pais "regiones Zingorum".[8] Et envoia le roy Alixandre

1168 Le] *om.* N; En enseust le B Le ... chapistre] *om.* HK .xxxviii.[e]] 38/38[me] MP; 36 D qui sont] *om.* N 1169 dehors] *om.* K; hors BGL climas] *add.* .xxxviii. K; *add.* .xxxviii. chap. H 1170 .iiii.] 4 MP; 2 D; quatre GH appert] fut dit K en] par FM 1171 ou] du BAF; de N .xxxi.[e]] 31[me] P; .xxi.[e] HKN; 21 M; 3 D lignes] *om.* K climat] *om.* K et] *om.* ECP 1172 .xii.] 12 DMP demi] ½ M Oultre] Decoste M; Jouste F les] *add.* autres K 1172-1173 aucteurs plusieurs] *om.* B 1173 de la est] *om.* D 1174 appellee] *om.* BHKN Taprobane] Aprobane K; Cryprobane B; probane *corr. to* Taprobane D; Traprobane HN a] *om.* M selonc] secun C 1175 Albategni] Abategni N; Albathegny L selonc] secun C car] que D 1176 habitans] habitations B leur] le M en] vers M et] *add.* soleil recouchant, c'est M 1177 a_2] par deca l'avons au M l'estoille] *om.* M 1178 en] *om.* M les estoilles] celles D 1179 en] *om.* B qui est] *om.* M est] *interlin.* C Canopos] Canepos ABFGHLMN; Canapos D selonc] secun C 1180 Solinus] Solinius HK est] *om.* GL ardant] *om.* K Albumazar] Albumasar FKN 1181 Zingorum] Rigorum HKN Alixandre] Alexandre ACDF

1 Diagram: Chapter 31 (after line 971).

2 "Taprobane" is modern-day Ceylon/Sri Lanka. To say that it is "opposite" India would be inaccurate, so the meaning must be "off the shore of" or "beyond", to the south.

3 See "Source References," line 1174.

4 Canopus (α Carinae, *olim* α Argus) is the second-brightest star in the southern night sky.

5 See "Source References," line 1179.

6 See "Source References," line 1180a.

7 See "Source References," line 1180b.

8 Land of the Zanğ or Azinge – often associated with Ethiopia. See "Source Reference," line 1181.

un legat en celle isle laquelle Plinius, Solinus, Pomponius[9] descripsent souffisaument et mettent moult d'autres isles et habitacions de celle part au dehors des climas.

Item: oultre la ligne qui termine les climas vers occident ha pou d'abitacions ou nulle, se ne sont, par aventure, aucunes petites isles, car illeques est la grant mer appelee "Oceanus".[10] Et la estoient, selonc ce que dit Haly, .iii. ymages ou ydoles en .iii. isles desquelles ymages chascune tenoit unes clefs en senefiant que en oultre n'a point d'abitacion.[11] Et dit Hely que elles sont pres du passage Hercules que nous appelons "les destres de Marroch". Et ces ydoles fist faire Hercules et sont appellees "Cades" ou "Gades" ou "les Columpnes Hercules".[12] Et dient aucuns que aussi mist il ydoles vers orient.

D'une de ces ydoles fait mencion Averoys en la fin du secont livre[13] *De celo et mundo*, ou comment. Et dit que elle estoit pres d'Espaigne et que il la vit toute droit eslevee et puis fu despeciee et destruite par les barbares qui orent guerre contre les

1182 en] q^r N Solinus] Solinius AH Pomponius] *om.* M; Pompinus G 1183 moult] *om.* D et$_2$] *add.* de GLN; *add.* des K au dehors] au dessoubz HKN; hors M 1184 oultre] *om.* D 1185 se] ce N; *add.* ce BK 1186 selonc] secun C .iii.] 3 P; trois BHM; 2 D 1187 .iii.] 3 P; trois BHMN; 2 D desquelles] desque K; lesquelles M desquelles ... chascune] et chacun de trois ymage B senefiant] signifiant BDFGLM; signifiance HK; segniefiance N en] *om.* DM 1188 dit] *om.* H Hely] Hali F; Haly CDGKLM; Hally H 1189 les ... appellees] *om.* N Marroch] Marrocht B; Maroth FG; Maroch KM; Meroot D 1190 Columpnes] *add.* de M 1192 ces] ses N ydoles] images K; *corr. from* ydoiles E Averoys] Averrois BH; Averroys MK; Avieis F du] de H 1192-1193 De celo et mundo] *om.* K; Du ciel et du monde FM; De celo et monde ADG 1193 ou comment] en arguante K ou ... que$_1$] et dit que au commancement M Et$_1$] il HN pres d'Espaigne] *rep.* D 1194 barbares] barbarins M

9 See "Source References," line 1182. The "legate" or messenger was Onesicritus.

10 The Atlantic Ocean.

11 See "Source References," lines 1186, 1188. 'Alī ibn Riḍwān writes that Hercules set up three statues or idols ("ymages ou ydoles") on three islands near the Strait of Gibraltar (Pillars of Hercules) to mark the passage.

12 Greek and Roman myths credit Hercules with opening up a passage between the Mediterranean Sea and the Atlantic Ocean as part of his tenth labour – the fetching of the cattle of Geryon. The mountains on each side of the Strait of Gibraltar became known as the "Pillars of Hercules" (or "Colonnes d'Hercule" in French), although the strait was sometimes also called the "Gates of Gades" (or "Cades" or modern-day Cadiz"). "Les destres de Marroch" ("les étroits de Maroc") would best be understood as "the narrows [i.e., strait] of Morocco", another name for this passage.

The "three idols" or "three images" could refer to some imagined physical pillars, north and south of the strait as well as at Cadiz. Or they could be the conversion of the idea of the "pillars of Hercules" into actual pillars. Latin versions of Aristotle use the phrase "idolum Herculis".

13 See "Source References," line 1192.

gens de celle isle en l'an de la predicacion de Mahommet .cccc. et .xxx. Ce fu environ l'an de l'Incarnation Notre Sires Jhesucrist, mil et .li.[14]

Item: oultre la ligne qui termine les climas vers septentrion ha moult d'abitacions comme Angleterre, Dace, Norvoie[15] et plusieurs autres regions, dont la desreniere, selonc aucuns, est une isle [a]ppellee "Tyle".[16]

Item: selonc Aristote et Averoys[17] en la fin du secont livre *De celo et mundo*, la fin de terre habitable vers orient, et la fin de terre habitable vers occident sont bien pres l'une de l'autre; et n'a entre .ii. fors que une mer qui n'est pas moult large. Et pour ce, en alant de l'une fin a l'autre, par terre habitable, ha plus d'espace grandement que n'est la moitié du circuite de la terre. Et donques se les climas se
f. 23^v traient en la / fin d'occident, si comme mettent les aucteurs, et il ne tiennent en long que la moitié du circuite de la terre, il s'ensuit, selonc Averoys,[18] que les climas ne se estendent pas siques en la fin d'orient, et que il ha grans habitacions oultre, hors

1195 de la] *om.* B; de FM Mahommet] Marhommet M .cccc. et .xxx.] .ccccxxx. L; 430 M; 830 D .xxx.] 30 P 1095-1096 la ... de] *om.* HKN 1196 l'Incarnation] *add.* de HN Sires] *om.* DK; Sire/Sue L; S^r MP; Seg~ B; Seigneur FH Jhesucrist] *om.* M; Ih'ucris E mil et .li.] mil .li. G; 1051 MNP; 105 D; .mli. FKN 1197 qui] que F 1198 comme] *add.* en HN Angleterre] *om.* B; Englcterre AFG 1199 selonc] secun C appellee] ppellee E Tyle] Tylle D; Tile G 1200 selonc] secun C et Averoys] *om.* F; et Averrois BH secont] 2 D mundo] mondo A 1202 l'une] une L .ii.] deux BDFHLKMNP fors] *om.* K que] *om.* FM pas] *om.* D 1203 de l'une] d'une ADFGLM a] en B l'autre] autre L 1204 n'est] ne ha F donques] pour ce D se_1] ses N 1205 traient] trennient K aucteurs] atteurs F; cosmographes M tiennent] trennent K 1206 selonc] secun C Averoys] Averrois BDFH climas] *marg.* C 1207 se estendent] d'estendent D siques] jusques DHKLM en] a KM que il] que illes M 1207-1208 oultre hors des] hors oultre les K

14 One of the three statues/idols (the one at Cadiz) reportedly still existed in Averroës' day, but during his lifetime it was destroyed when Berbers/ barbarians invaded the island (ibid.).

The dates given are problematic. First the year 430 AH converts to the Julian year 1039, while the Julian year 1051 converts to 443 AH. Second, the date given in the critical edition of Averroës' commentary is "530" (ccccc xxx); "430" appears only in a marginal note to one manuscript. Third, if, as Averroës indicates, the statue was destroyed during his lifetime, the event would have to have happened between 520 AH and 595 AH. Fourth, in the *Livre du ciel et du monde*, Oresme (p. 567) gives the date 430 AH, but relates this to 1057 AD.

15 England, Denmark, Norway.

16 Thule – the most northerly location in ancient Greek and Roman literature, sometimes identified with the islands north of mainland Scotland. In the middle ages it was also used metaphorically to mean any place beyond (i.e., north of) the known world.

17 See "Source References," line 1200.

18 A source for this concept in Averroës has not been identified.

des climas par devers orient, ou il couvient que les climas, ou aucuns d'iceuls, soient plus longs que les astrologiens ne mettent.

1208 des] les P par devers] jusques en la fin de M ou_1] *illeg.* D soient] sont K

LE .XXXIXE CHAPISSRE. D'UNE MERVEILLEUSE CONSIDERACION OU CIRCUITE DE LA TERRE

Poson que selonc la longitude des climaz l'en peust circuier ou environner ou aler tout entour la terre en alant tout droit devant soy vers orient ou vers occident, tant que l'en fust revenu au lieu dont l'en se seroit parti premierement.

Et, selonc Aristote et Aveuroys, il s'en faut pou que l'en ne puet faire cil tour ou cil pelerinage, si comme il fu dit au chapistre precedent.[1]

Or posons donques pour cause d'exemple que .i. homme ou .i. oiseau feist cil tour en .xii. jours naturels en alant regulierement vers occident, et commencast maintenant a midi.

Il appert clerement que il passeroit en chascun jour naturel la douzieme partie du circuite de la terre, ce sont .xxx. degrés.

Donques couvendroit il que le soleil feist .i. tour environ la terre et .xxx. degrés oultre avant qu'il retournast landemain au meridian de cil homme. Et ainsi auroit icil homme son jour et nuit de .xxvi. heures, et plus long par la douzieme partie d'un jour naturel que se il se reposast. Par quoy il s'ensuit de necessité que en ces .xii. jours naturelx cil homme auroit tant seulement .xi. jours et .xi. nuis, et que le soleil

1210 Le ... chapistre] *om.* HK .xxxixe] 39/39me DMP merveilleuse] *add.* et belle M 1211 ou circuite] *om.* N terre] *add.* .xxxix. HKN; *add.* Le chappitre des causes par quoy aucun lieu est habitable ou non D[2] 1212 selonc] secun C l'en] le L; on P; l'an G ou$_1$] et M environner ou] *om.* L 1213 la ... tout$_2$] *om.* FM 1214 l'en$_{1,2}$] l'an G se] *om.* BCDG seroit] soit L premierement] pre D 1215 selonc] secun C Aveuroys] Averrois BDHM; Aurois F il] *new paragraph actually begins here in* E l'en] l'an G ne puet faire] feront B cil] le M 1216 ou cil pelerinage] *om.* M cil] *om.* B 1217 que .i.] d'ung P .i.$_{1,2}$] un/ung ABDFLMNP 1218 .xii.] 12 DM regulierement] regleement M; reguleement C 1219 maintenant] *om.* D 1220 douzieme] 12 D; .xii./.xiie BFGKMN 1221 .xxx.] 30 MP 1222 Donques ... degres] *om.* *K*M .i.] un/ung ABDFHLNP .xxx.] 30 DP 1222-1223 et ... oultre] et en eoultre qu'il passast .xxx. degrés B 1223 retournast] tournast FM cil] cest M; icellui P cil homme] cellui DK icil] icellui P; ycelui L; cest M 1224 et nuit] *om.* M .xxvi.] 26 DMP et$_2$] *add.* ainsi/aussi seroit ADFGKLMN; *add.* ainsi seroit il B par] de GL douzieme] .xiie BGHKLN; 12/12e DM d'un] du BFM 1225 naturel] *om.* B se il se] cil s'en K; cil se HN se$_2$] *om.* F .xii.] 12 MP;.xi. L 1226 cil] cellui M; icellui P aurait] *add.* fat G; *add.* fait L .xi.$_{1,2}$] 11 DMP

---

1 Chap. 38, ll. 1200-1209. The sources in Aristotle and Averroës are the same as for line 1200.

2 This is actually the title for Chap. 40.

ne li leveroit que .xi. fois, ne ne reconseroit que .xi. fois, quar .xi. jours et nuis, chascun jour et nuit de .xxvi. heures, font .xii. jours naturels, chascun de .xxiiii. heures.

Item: puis que il passe .i. jour naturel la douzieme partie du circuite de la terre, et son jour et nuit est plus long. ii. heures que le jour naturel, il s'ensuit que en son jour et nuit il passe plus que la douzieme partie du circuite de la terre, et que il passe .xxxii. degrés et demi du circuite de la terre. Je appelle son jour et nuit le temps que le soleil met a retourner de son meridian d'un jour a son meridian de landemain, ou de son soleil levant a son soleil levant.[3]

Item: par semblable consideracion, couvendroit il que .i. autre homme qui /
f. 24r feroit ce tour en alant droit vers orient eust son jour et nuit plus court ou plus brief que n'est .i. jour naturel. Et donques s'il faisoit cil tour en icelui meisme temps, c'est assavoir en .xii. jours naturels, il s'ensuiroit par necessité qu'il eust en cil temps

1227 .xi.$_{1,2,3}$] 11 DMP ne ne] ne liu F; ne aussi ne lui M; *add.* lui/luy ABD ne ne ... fois] *om.* GL quar] $\overline{qr}$ ECBD; car PAFGHKLMN 1228 .xxvi.] 26 DMP .xii.] 12 DMP .xxiiii.] 24 DMP; .xiiii. HN; .xxiiii. *corr. to* .xiiii. K 1228-1229 chascun ... heures] *om.* F 1230 .i.] un/ung ALP; en un BDHKMN; en .i. C naturel] *om.* K; *add.* il s'ensuit D douzieme] 23[e] M; 12[e] D; .xii.[e] BFHGKLN du circuite] *om.* B 1231 .ii.] 2 DP; deux FN; de 2 M .ii. heures] *om.* B 1231-1232 en .. nuit] *om.* GLM 1232 il$_1$] *interlin.* C que$_1$] de KM douzieme] .xii.[e] GHKLN; 12[me] DMP du circuite] *om.* B 1232-1233 et ... terre] *om.* N 1233 .xxxii.] 32 DMP demi] ½ M de la terre] d'icelle M Je] Il K 1234 d'un ... meridian$_2$] *om.* D jour] *add.* jusques B 1234-1325 ou ... levant] *om.* B 1235 de] du M a son] au M son$_2$] *om.* D levant] couchant AFGHLN; *add.* de/du landemain KM 1236 .i.] un/ung ABDFHLKMNP autre] *om.* GL 1237 droit] *om.* GLM ou] et M 1238 .i.] un/une/ung ABDFHLMP cil] icellui P icelui] cellui M 1239 en$_1$] *om.* D .xii.] 12 DMP

3 While our traveler (he will be called Jehan – line 1247) moves what he thinks is 30 degrees (from noon on the first day to "noon" on the second), the sun moves around to its position at the first noon and then continues for a little more than another 2 hours to reach the man's new "noon position" on the second day. (Actually, the sun advances an extra 32°43′ – i.e., 360° plus 32°43′ or 392°43′–because in the meantime, the man has moved on – and carried his "noon" forward as well.) Thus when the traveller circles the earth in eleven 26-hour "days" of 32°43′ each (during each of which he covers 1/11th of the circumference instead of 1/12th), the sun will have reached him at "noon-time" (i.e., when it is directly over his head) 11 times, but at the final noon the sun will have reached the first position (from whence the traveller set out) – in the eyes of someone who remained there – 12 times – 12 (24-hour) days.

This is the phenomenon which is the basis of Jules Verne's 1872 novel, *Around the World in Eighty Days* (*Le Tour du monde en quatre-vingts jours*), although Verne's protagonist (Phileas Fogg) travels eastwards. The effect of travelling east will be discussed by Oresme in the following lines.

meismes .xiii. jours et .xiii. nuis, et seroit son [jour et] nuit ensemble de .xxii. heures et .ii. tresiemes d'une heure – ce sont .xxii. heures et pres que .ix. minus.

Item: puisque cil homme passe en un jour naturel la douzieme partie du circuite de la terre, et son jour et nuit est plus court que du jour naturel, il s'ensuit que en son jour et nuit, il passe moins que la douzieme partie du circuite de la terre, et que il passe precisement .xxvii. degrés et .ix. tresimes d'un degré – ce sont aussi comme .xxvii. degrés et .xlii. minus du circuite de la terre.[4]

Je di donques que se Jehan faisoit cil tour en alant vers occident, et Pierres le feist en alant vers orient, et partissent l'un quant l'autre, et revenissent ou acomplissent l'un quant l'autre, Pierres auroit en cil meismes temps .ii. jours et nuis plus que Jehan. Et Robert, qui se reposeroit, auroit .i. jour plus que Jehan, et .i. jour moins que Pierres,[5] combien que ils meissent a faire cil tour, et fust .xii. jours ou cent ou .x. ans, ou combien que il y meissent.

1241, 1246 minus] *read* minutes

1240 meismes] *om.* L .xiii.$_{1,2}$] 13 DMP jour et] *om.* E; *interlin.* C nuit] *add.* tout GL .xxii.] .xxiii. A; 22 DMP 1241 .ii. tresiemes] 2/13 P; deux .xiii.^{es} B; .ii. .xiiieme HK; 2 tresiemes D .ii. ... et$_2$] *om.* MN d'une heure] *om.* K; d'une H .xxii.] 22 P; 12 D que] de M .ix.] 9 DMP 1242 cil] cest M; icellui P un] .i. G naturel] *om.* N douzieme] xii^{e} ABGHKLN; 12me DMP 1243 du] un/ung CPABDFHLM; .i. G en] *om.* D 1243-1244 et$_1$ ... terre] *om.* KN 1244 que$_1$] de M la$_1$] *om.* G douzieme] xii^{e} BFGHL; 12me DMP du circuite] *om.* B il$_2$] *interlin.* C 1245 .xxvii.] 27 DMP .ix. tresimes] 9/13 M; 11/13 P; .ix. .xiii.^{es} BHKN; 9 tresimes D d'un] de M d'un degre] *om.* D aussi] ainsi M 1246 .xxvii.] 27 DMP .xlii.] 42 DMP 1247 cil] icellui P 1548-1249 ou acomplissent] en acomplissant leur tour M 1249 quant] comme D cil] icellui P meismes] *om.* B .ii.] 2 DM; deux BP jours] *interlin.* A et] *add.* .ii. AGKN; *add.* 2 DM 1250 Et$_1$ ... Jehan$_2$] *om.* N .i.$_1$] *om.* D; un/ung ABFHLMLP .i.$_2$] *om.* D; un/ung ABFGHLKMNP jour$_2$] *om.* B 1251 cil] icellui P fust] pose que il y meissent ABDFGHKLN .xii.] 12 DP 1251-1252 et fust ... meissent] ou 12 jours ou 10 ou 100 M 1252 cent] .c. FK; 100 D .x.] 10 DP; dix B

4 Pierre (line 1247) travels eastward, and although he thinks he has travelled 30° each day ("noon" to "noon"), in fact he has only travelled a little over 27° 42′ each day, so that it takes him 13 of these shorter "days" (of 22 hours 9 minutes) to circle the earth, although only 12 days have passed at the starting/ending point itself.

5 Thus when Jehan and Pierre reach their starting point – where Robert has been sitting in a taverne all this time – it appears that Pierre travelled 2 days longer than Jehan – at least he has seen 2 more sunrises and sunsets than Jehan. But Robert finds himself counting one day more than Jehan and one day less than Pierre. (Of course, after 12 days in a taverne it is debatable that Robert had any idea what day it was!)

And, as Oresme continues, if Jehan and Pierre left Paris on a Tuesday and arrive back on a Sunday, Jehan thinks that it is Saturday and Pierre thinks it is Monday. (Either way, they are both probably late for Mass.)

Item: il est bel a considerer comme Jehan et Pierres revendroient ou arriveroient a .i. meismes jour aussi comme au dyemenche pour cause d'example; et nientmoins Jehan diroit que il est samedi, et Pierres diroit qu'il est lundi. Et y auroit grant consideracion et de l'assignacion des jours quant a l'appellacion, et de ce que ce est pou de chose ou nient de la dominacion des planetes selonc les heures du jour et d'autres choses que je lesse pour cause de brieté.[6]

1253 Item] *om.* B Item il] Plus est M il] *om.* KN comme] comment GL; que K et] ou B arriveroient] environneroyent K 1254 a .i.] en un/ung FLM, a un/ung ABDHKNP; en .i. G aussi] ainsi M 1255 Jehan diroit] *rep.* P est$_1$] seroit B; yert H est$_2$] seroit B y] il D auroit] *add.* il B 1256 de l'assignacion] grant differance B; la significacion HN; la de signacion *corr. to* de l'asignacion L ce$_2$] *om.* GLM 1257 dominacion] duātion/dnātion A selonc] secun C 1258 causee de] la K breite] brefveté M

6 Another way of looking at this is to consider the total period during which the two men are away – 24 hours times 12 days, or 288 hours – and understand that Jehan, heading west, divides this period into 11 units during each of which he travels more than 30°, while Pierre, heading east, divides this period into 13 units during each of which he travels less than 30°. Their error is in counting each of these (11 or 13) units as a "day" when in fact they are either more than a standard 24-hour day or less depending on whether the sun has to catch up to each man's new "noon", or meets the new "noon" in less than 24 hours.

DIAGRAM: CHAPTER 39

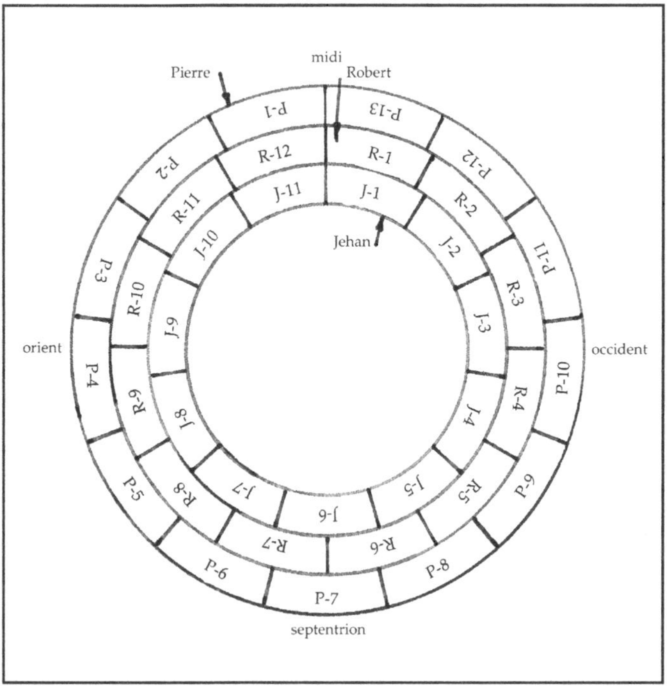

La figure d'une consideration du circuite de la terre

Complete diagram: E CP
Alternative diagram: BFHKN
No diagram (or space only): DGLM

DIAGRAM: CHAPTER 39 ALTERNATIVE

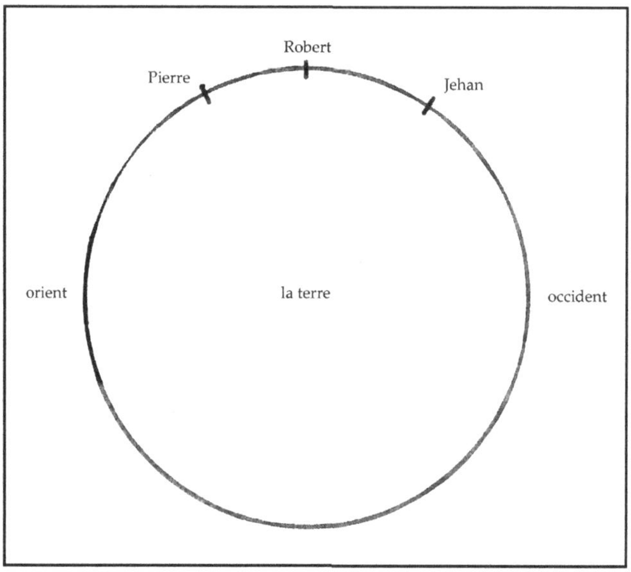

Complete diagram: BFHKN

Robert] R HK Pierre] P HK Jehan] J HK orient] *om.* K occident] *om.* K

f. 24v /

LE .XLE CHAPISTRE. DES CAUSES POURQUOY AUCUN LIEU EST HABITABLE OU NON

Les medicins dient que, quant asavoir la nature de corps humain, il couvient resgarder a la racine de dessus, c'est asavoir au ciel et sa disposition, et si couvient resgarder a la racine de dessous, c'est asavoir a la complexion ou la disposition de la personne.

Semblablement, quant a l'abitacion de la terre, une cause general est au regart du ciel, c'est asavoir moderee ou attrempee distance de la voie du soleil, car estre en trop pres ou trop long est cause, quant est de soi, de trop chaut ou de trop froit, si comme il fu dit ou .xxviii.^{e} chapistre.[1] Mais autres causes especiaux pueent estre au regart de la terre. Et sont .iii. principalment qui font aucune partie de la terre estre inhabitable en distance couvenable de la voie du soleil et dedens les climaz.

f. 25r Une est quant celle partie est couverte de mer ou empechíes / par eaue; l'autre est quant elle est sterile et ne porte bon fruit, ou occupee de serpens ou de mauveses bestes, si comme sont aucuns desers.

La tierce est la disposicion, le siege, la figure d'aucuns lieux, montaignes ou valees qui sont inhabitables pour trop froit ou trop chaut ou pour trop soutil aer ou trop gros, ou pour autre disposition de l'aer, qui n'est pas couvenable a nature humaine.

1259 Le ... chapistre] *om.* HKN .xl.^{e}] 40/40me DMP Des causes] *om.* D aucun] un/ung BFHKMN 1260 ou non] *om.* N; *add.* .xl. KN; *add.* .xl. chap. H; *add.* Des extremites de terre habitable H[2] 1261 quant assavoir] pour savoir M de] des FKM; du GL 1262 de] *interlin.* L; des N ciel] thies(?) D et$_1$] a M; et a DN sa] a la K 1263 de$_1$] *om.* B a$_2$] *om.* L complexion ... disposition] disposition et (a) la complexion HKN ou] a GL; ou a FK 1265 la] *om.* K au] du F 1266 voie] *add.* du ciel ou HKN car] par HKN en] *om.* HKN 1266-1267 estre ... long] la propinquité ou loingtaineté M 1267 quant ... soi] *struck out* D 1268 .xxviiie] 28/28me DMP causes] *add.* tantot P 1269 terre] *add.* habitable M sont] fut F .iii.] 3 DP; trois BH la] *om.* HKN estre] *om.* BDFGHKLN; *interlin.* C 1269-1270 Et ... inhabitable] *om.* M 1270 en] ou K 1272 est$_1$] *om.* ABDFGHKLM 1273 aucuns] autres K; *add. and del.* mauvais C desers] dessus K; dessus dites N 1274 tierce] 3 D le siege] le signe et M; du sterle D la figure] *marg.* G 1275 qui sont] *om.* L froit ... chaut] chault ou trop froit M trop$_2$] *om.* D; pour N; pour trop K trop$_3$] *om.* N ou$_3$] *add.* pour K 1276 autre disposicion] autres disposiions M

---

1 Chap. 28, ll. 827-832.

2 This is from the title to Chap. 41.

Item: en distance desattrempee de la voie du soleil, et qui seroit inhabitable quant est de soi, il puet estre que .iii. causes contraires et opposites as .iii. causes dessus ditez couvienent ou con[c]uerent ensemble, et pour ce, est tel lieu habitable. Et ainsi dient les hystoriens que les montaignes Yparborees[3] sont de grant attrempement, et y vivent les gens tres longuement et aucuns tant que il les couvient tuer. Et toutesvoies, il sont hors et loing des climaz vers septentrion. Et Lincoliense[4] declaire la cause et dit que ce est pour la figure et la disposition de ces montaignes, car il ont la superfice, ou la partie qui est devers le soleil, planee ou bien polie et concave. Par quoy il recoivent et retornent convenablement la lumiere et la chaleur du soleil, et portent ombre ou abri, et deffendent les habitans du vent de bise. Et pour ces .ii. causes n'i a il pas grant froit.

1278 en] en la D 1279 .iii.$_{1}$] 3 D; trois/troys BHMNP .iii.$_{2}$] 3 D; trois/troys BFHMNP 1280 concuerent] contuerent E et pour ce est] avant que soit D(*later hand*) tel] ce HKMN; cest AFGL; la D 1281 ainsi] aussi M 1282 tres longuement] trop grandement B aucuns] aulouns y vivent M les$_{2}$] *om.* B 1283 hors et loing] loing et hors F et loing] *om.* A des] les L Lincoliense] *added later* A; Linconiese DF; Lencolieuse K 1283-1284 Lincoliense ... cause] *om.* M 1284 la cause] *om.* K dit] disent M la$_{3}$] *om.* BKM 1285 car] que D ou$_{2}$] et BP 1286 il] *om.* M recoivent et retornent] couvient tretornent L; recouvient tretornant G; recouvinet et retournient K recoivent ... couvenablement] recepvent modelment(?) B retornent] recouvrent M lumiere] *add.* du soleil FM 1287 portent] *add.* celles montaignes M 1288 .ii.] deux BHMP il] *om.* B; 2 D(*interlin.*)

3 Traditionally "Hyperboreans" refers to people, and not mountains. They supposedly lived far to the north, beyond the normal habitable lands, in an area thought to be sunny and temperate because it was protected by mountains – generally identified as the Riphean Mountains (see below, the appendix on "Climes"). The earliest extant reference is in Herodotus, *History*, Bk IV, cap. 32-36.

4 See "Source References," line 1283.

Item: se les causes especials de bonne habitacion couvenoient et concueroient aveques la general, c'est assavoir terre bien fertile et bien assize a souhet aveques bon regart du ciel, adonques tele region seroit tres attrempee. Et, par aventure, tel est le lieu que les poetez, selonc Aristote,[5] appellent les "ysles beneurees",[6] et puet estre que tel est paradis terrestre.

1289 Item] Et L se] *om.* M habitacion] *add.* si M 1290 a] et F 1291 tele] celle H seroit] *om.* M tres] bien BD; *add.* bien L(*interlin.*)G 1292 selonc] secun C selonc Aristote] *om.* B 1293 que tel est] *om.* B terrestre] *add.* la ou Adam et Eve estoient immortals D(*later hand*)

5 See "Source References," line 1292.

6 The "Islands of the Blessed" is from (at least) Hesiod (*Works and Days*, ll. 156-170), and they are supposedly where the heros of Greek mythology (as, e.g., in Homer) were settled by Zeus – "apart from human kind, at the end of the world." Aristotle at least hedges his bets by referring to these persons (i.e., the heros) with the disclaimer "if such there be".

Some identified the Islands of the Blessed with the Elysian Fields, and placed them in the Antipodes (John Kirland Wright, *The Geographical Lore of the Time of the Crusades* [New York: American Geographical Society, 1925 / Dover Reprint, 1995] p. 28). In the middle ages it was also sometimes identified with the Garden of Eden.

LE .XLI.E CHAPISTRE. DES EXTREMITÉS DE TERRE HABITABLE

Ces .iiii. lignes qui [sont] termes et extremités de terre habitable furent touchiés ou .xxxviii.e chapistre.[1] Et l'extremité devers midi et celles devers septentrion sont desattrempees, si comme je ai dit souvant.[2] Et, pour ce, dit Ptholomee en son *Quadriparti*[3] et Haly[4] et les anciens que en ces .ii. extremités ha hommes sauvages qui menuient les gens, et qui ont les visages et les faces corrompués et contrefaites. Et la cause si est, selon Haly,[5] pour la desattrempance de ces regions en chault et en froit, par quoy les corps sont de complexion desordenee et de mauvese et lede composition. Et, pour ce, sont il de mauveses meurs et de conversation estrange et f. 25v sauvage. Et sont illeques, comme dit est, gens de horrible figure et monstres / ou bestes en tant que a paine puet on cognoistre se ce sont hommes ou non, si comme recite Saint Augustin.[6]

1294 Le ... chapistre] *om.* HKN .xli.e] 41/41meDMP chapistre] *om.* G de] de la G habitable] *add.* et differance de leur latitude D; *add.* .xli. chap. HN; *add.* .xli. K 1295 Ces] Les AGKLMN .iiii.] 4 P; quatre BHMN lignes] sigñ N sont] *om.* ECP termes et] les GL 1296 .xxxviii.e] 38me P; 36/38 D; 28 M Et$_1$] *add.* de D et$_2$] *add.* de D septentrion] *corr. inter. from* atrion D 1297 Ptholomee] Ptholome B; Ptolomee CF 1298 Quadriparti] Quadripartis HN et$_1$] que D anciens] aultres KM ces] ses N .ii.] 2 DP; deux BHMN 1299 qui] que B menuient] mangent B; menguent HN et les faces] *marg.* F corrompues et] *om.* GL 1300 si] *om.* ABDFGHKLMN selonc] secun C Haly] *add.* Haitibagel(?) N pour] *om.* D desattrempance] destrempance H de ces] des K 1301 par] pour HKN quoy] *om.* GL desordenee] *om.* B 1302 mauveses] *add.* et laydes M 1303 comme dit est] *om.* ADFGLM gens] *om.* B ou] et aultres M 1304 tant... peut] telle maniere peust M a ... on] apoy l'en peut B a paine] a poyne / a poy ne ADFGHN; a pou ne KL on] l'en HKN se] ou B se ce] s'ilz HKN ce] *om.* GL

1 Chap. 38, ll. 1170-1173.
2 Chap. 28, ll. 825-827; cap. 30, ll. 914 -916; cap. 31, ll. 944-949.
3 See "Source References," line 1298a.
4 See "Source References," lines 1298b, 1300,
5 Ibid.
6 See "Source References," line 1305.

Item: Haly[7] dit en especial de ceuls devers midi qui il sont vers et creps fors[8] sous l'equinocial ou il sont geaunes, car il y a plus grant attrempement, si comme il fu dit ou .xxixe chapistre.[9]

Item: Haly recite que Hermes[10] dit que es .ii. extremités dessus dites habitent mauves esperis et dyables et bestes malfaisantes as hommes. Et est assavoir que les .ii. autres extremités de habitacion, c'est celle d'orient et celle d'occident, ne sont pas ainsi desattrempees fors tant comme il participent aveques les autres .ii., c'est asavoir es .iiii. angles de terre habitable, desquels je fis mencion ou .xxxi. chapistre,[11] car ces .ii. extremités es autres parties sont en moderee et attrempee distance de la voie du soleil, si comme il appert clerement par le .xxxie chapistre.[12]

1306 de] que N midi] *add.* que il ABCDFGMP sont] *om.* B; qu'ilz L vers] *om.* D; noirs AFGHKL; *add.* .i. nigre N creps] tres ABDFHKMN creps fors] froiz GL 1307 geaunes] jausnes B; geamies/geannes L; janues/jamies HN car] que D 1308 .xxix.^{e}] 29 DMP 1309 recite] dit HK; dist N .ii.] 2 D; deux BHKMN; *add.* aultres M 1310 et dyables] *om.* FM as] a H as hommes] *om.* K 1311 .ii.] 2 DP; deux BHMK de] *add.* terre L occident] *add.* mais elles M 1312 comme il] qu'ilz DLM; qu'elles HN .ii.] 2 D; deux BHMNP 1313 es] avecques les K .iiii.] 4 DMP; troys B .xxxi.] 31/31me MP; 3 D; .xxi. GL 1314 .ii.] deux BHMNP autres] .ii. L en] *om.* L; a DHKN et] et en B et attrempee] *om.* C 1315 si] *om.* HN clerement] clerent E .xxxi.^{e}] 31/31me MP; 3^{e} D

7 See "Source References," line 1306.

8 A slightly convoluted excerpt from Ptolemy and 'Alī ibn Riḍwān who originally were writing about the people – their bodies and their hair – but here the phrase seems to concentrate on only the hair. "Vers" would be "wormy" – the body is infected with worms, or the hair looks like worms. "Creps" is related to "créper" or "crépir" (both from the same root), the first meaning "to crimp" hair, the second "to roughcast" as to roughly plaster a wall. Both may also describe the texture of tough hide or skin. "Fors" is probably "forts", that is, "very crimped." Later versions present this as "black and very strong" which, while not the original, also makes sense.

The original Latin describes the hair as "black, thick, and crinkly."

9 Chap. 29, ll. 827-829.

10 See "Source References," line 1309.

11 Chap. 31, ll. 955-957.

12 Chap. 31, ll. 958-962? The reference is unclear.

LE .XLIIE CHAPISTRE. DE LA DIFFERENCE DES PARTIES BIEN HABITABLES SELONC LEUR LATITUDE

En terre habitable, aillours que es extremités dessus dites, ha difference selonc longitude et selonc latitude, mes la plus grant difference est selonc la latitude.

Car ceulz qui sont plus vers midi, si comme dit Aristote ou .viie livre *De politiques,*[1] sont de plus grant entendement et de plus grant prudence et mains courageuz, mains fors ou mains hardiz. Et pour ce sont il subjects anciennement. Et au contraire ceulz qui habitent plus vers septentrion sont plus fors, plus hardis et plus courageuz, et de mendre prudences. Et se tiennent et gardent en leur libertés.

Et dit Aristotes[2] que les Grecs sont moiens et ont souffisaument force et prudence. Et pour ce, dit il que il sont, selonc nature, habiles a avoir seignourie sur les autres qui sont plus vers midi ou plus vers septentrion. Mes il me semble que les Grecs n'ont pas eue continuelment tele seignourie, mes fu transferés as Rommains en Ytalie, selonc les hystoires. Et cuide que aveques la cause dessus dite de maintenir roial seignourie et bonne policie, sont autrez causes ou ciel qui sont aucunes fois plus fortes, si comme, par aventure, le mouvement de *aux solis* et de la huitisme espere, ou aucun autre mouvement qui puet estre cause de envillissement d'un royaume, et de deffaute de bon gouvernement, et de conversion a tyrannie, par f. 26^{r} quoy il s'ensuit dissipation, division ou transla/tion de seignourie. Et n'est pas mon intencion que ce soit cause necessaire, mes tant seulement comme le corps du ciel nous enclinent a diverses choses.

1316 Le ... chapistre] *om.* HNK .xlii.^{e}] quarante deusieme L; 42/42me DMP 1317 habitables] *add.* .xlii. K; *add.* .xlii. chapistre HN selonc] secun C selonc ... latitude] *om.* HN latitude] longitude D 1318 es] *om.* N; *add.* autres GL selonc] secun C 1319 selonc$_{1,2}$] secun C la$_{1}$] *om.* DFGKLN 1320 vers] devers HN .vii.^{e}] septieme FGL; 7^{e} DM; 8^{e} P livre] *om.* L 1321-1322 mains courageuz] *om.* K 1322 sont] fiuent K; fure(?) D anciennement] *om.* K; entiannement B; aultrinnement M 1323 plus$_{1}$] *om.* G 1324 et$_{1}$ ... prudences] *om.* B 1325 et$_{2}$] car ils M 1326 dit il] *om.* D selonc] secun C selonc nature] *om.* B 1326-1327 sur ... septentrion] *om.* F 1327 plus$_{2}$] *om.* L 1328 Rommains] Romains L 1329Ytalie] Ytaillie K selonc] secun C cuide] en iude B que] *add. and del.* aucuns E aveques] *om.* D; oncques FGHLMN 1330 roial] a royal HK 1332 la huitisme] la 8^{e} DM; le 8me P; le .viii.^{e} BHKN; l'octisme L; le octisme AFG ou] *om.* D ou ... mouvement] *om.* F aucun] *om.* M 1333 d'un] du AFGL; de BDHKMN de deffaute] de faute D bon] bien F conversion] *add.* .i. conjunction N 1334 s'ensuit] *add.* la M dissipation] disposition ABDFGHKLMN; *add.* a DFGL A(*interlin.*); *add.* et M pas] *om.* D 1335 tant seulement] que tant D comme] *om.* B

1 See "Source References," lines 1321, 1325.

2 See "Source References," lines 1321, 1325.

Item: Ptholomee, en son *Quadriparti*,[3] dist semblablement que ceuls qui habitent plus vers midi sont de plus grant entendement et plus habiles et plus enclins, et de plus grant vertu a savoir les sciences des estoiles et les ars mathematiques. Et dit Haly[4] que les plus sages ont este du tiers, du quart, et du quint climaz; et que Aristote et Galien furent du quint; et Ptholomee et Plinius furent du tiers.[5] Et est ceste regle ou difference dessus dite a entendre quant a plusieurs, et communelment ou selonc commun cours.

1337 Ptholomee] Ptholome B; Ptolomee F Quadriparti] Quadriptit M semblablement] *om.* M 1338 plus$_1$] *om.* K et$_1$] ou ABDGHK et$_2$] ou B 1339 les$_2$] des K 1340 Haly] Hali F tiers] 3 D quart] 4 D quint] 5 D que$_2$] *om.* P 1341 quint] 5 D; *add.* climat M Ptholomee] Tholome B; Ptolomee F tiers] 3 D 1342 ou] en BH entendre] attendre B 1343 ou selonc] ausegon B; ou second HN; ou secun C

3 See "Source References," line 1337.

4 See "Source References," line 1340.

5 The point being made by these examples is unclear. Aristotle was from Stagiria (northern Greece; 40°36′N) and Galen was from Pergamon (now Bergama, western Turkey; 39°08′N) which puts them in the 4th clime and not the 5th. Ptolemy was from further south (born possibly in middle Egypt but living mostly in Alexandria; 31°12′N) which is in the 3rd clime; but Pliny the Elder was from northern Italy (Como in Lombardy; 45°49′N) which is in the 6th clime.

LE .XLIII.[E] CHAPISTRE. DE LA DIFFERENCE DES PARTIES BIEN HABITABLES SELONC LEUR LONGITUDE

L'autre difference des parties bien habitables est selonc la longitude des climaz, et n'est pas si grande ne si notable comme la difference de la latitude. Et de ce dit Ptholomee[1] que ceulz devers orient sont plus masculins et de plus fort cuer et monstrent tout leur courage. Et dit Haly[2] que euls ne scevent leur fais en rien celer ne couvrir et ont plus grant puissance de faire choses merveilleuses et de savoir astrologie, car il sont en orient qui est la dextre partie du monde, et sont plus de la nature du soleil.

Et, au contraire, la partie devers occident est de la nature de la lune, et ceulx de ces parties, celent leur courage, et sont contencieus a guise de fames.

Et quant ils ont victoire sus leur adversaires, il les occient et les tuent. Et ne le font fors par couardise et pour la grant paour que il ont que autrefois ne se esdressacent sus euls et que il ne se revengassent.

Et adjouste Haly[3] que ceci est a entendre quant a plusieurs, quar aucune fois seront .ii. hommes, l'un d'orient, l'autre d'occident, et toutesvoies celui d'occident sera de plus fort cuer que l'autre. Et ainsi puet il estre des autres conditions dessus

1354 fames] *read* femmes

1344 Le ... chapistre] *om.* HKN .xliii.[e]] $43/43^{me}$ DMP chapistre] *om.* G 1344-1345 De ... longitude] Du croissement et appetissement de la lune D[4] 1345 bien ... longitude] *om.* N selonc] secun C longitude] *add.* .xliii. K; *add.* .xliii. chap. HN 1346 selonc] secun C la] leur B 1347 de_1] prise selon M la_2] *om.* DFK 1348 Ptholomee] Ptholome B; Ptolomee F ceulz] *add.* de P $plus_2$] *add.* grant M cuer et] en M 1349 Haly] Hali F leur ... rien] en rien leurs faiz M celer] sceller B 1351 car] quar B; que D il] ilz PDHKLMN 1353 la partie] sont ceulx M occident] *add.* car icelle partie M 1354 ces] ses N celent] selon ABDGHKLMN et sont] *om.* K contencieus] courreceux L a guise] et de manieres M fames] feme(s)/femme(s) DFGHKLMNP: fõmes A 1355 les occient et] *om.* M les_1] *om.* F 1356 fors] *add.* que MP la] *add.* couardise et B autrefois] *add.* il/ils ABDHKLMN 1357 esdressacent] adressassent DP; adrocassent K; esdrecassent HN; esdrescassent AL; esmeuvent M; revangeassent B sus] contre ADFGHLMN et ... revengassent] *om.* B il] *om.* D revengassent] revenchassent K; revengeassent AF; revengent M; revengeussent P 1358 Haly] Hali F entendre] attendre B quar] car PAFGHKLMN; que D 1359 seront] seroient F .ii.] 2 D; deux BGHNP toutesvoies] dont M 1360 sera] *add.* plus courageux et M puet il estre] est il a autendre B

---

1 See "Source References," line 1348.

2 See "Source References," line 1349.

3 See "Source References," line 1358.

4 Copied from Chap. 44.

dites, mes il n'avient pas communelment. Et par ce appert que ceulx de ces pais de France ou de ces parties ne sont pas communelment habiles a savoir astrolgie, especiaument de jugemens, car il sont vers septentrion et vers occident, et, selonc Ptholomee en son *Centiloge*[5] et selon raison, ceulz y pueent peu profiter qui n'i sont enclins de nature. Et pour ce, selonc astrologie meiwme il appert que ceulz de ces
f. 26v pais ne pueent pas / communelment profiter en astrologie judicative, et doit on pou de foy adjouster a leur jugemens meesmement quant as effis de fortune. Et se ai je declairié plus plainement en .i. traitié[6] que je ai fait de ceste matiere.

1361 n'avient] ne advient MN communelment] continuellement M ce] cest (*interlin.*) L; *add.* il B de ces] du B 1362 ces] ses N pas] *om.* B communelment] comement L 1363 especiaument] espalment G de] en HKN; des M car] que D selonc] secun C 1364 Ptholomee] Tholome B; Ptolomee F Centiloge] Centiloque M et ... raison] *om.* B selon] secun C y] n'i B peu] *add.* petit M qui] si il B n'i] ne N 1365 selonc] secun C meisme] mesmement HN 1366 on] *om.* GL; l'en B 1367 a] en HKN se] *om.* B; ce AN je] *om.* H 1368 plus] *add.* a F plainement] a plain M .i.] un/ung ABGHKLNP; *add.* autre B

5 See "Source References," line 1364.

6 *Livre de divinacions*. See G. W. Coopland, *Nicole Oresme and the Astrologers. A Study of His Livre de divinacions* (Cambridge MA: Harvard University Press, 1952). This possibly could also be a reference to Oresme's Latin treatise against astrology: *Tractatus contra astronomes judiciarios*.The language of Oresme's text is not specified here, unlike in the reference in Chap. 50.

LE .XLIIIIE CHAPISTRE. DU CROISSE[ME]NT ET APPETISSEMENT DE LA LUNE

Aprés ce que j'ai dit et parlé de la terre, je vueil retourner a parler du ciel et dire les causes des eclipses, mes je vueil avant declairer comme la lune est enluminee. Et est asavoir que elle n'a point de lumiere quant est de soi, mes est illuminee du soleil, et ainsi sont toutes les autres estoiles, selonc ce que dit Averroys.[1] Et est la lune roonde et sperique a maniere d'une pomme ou d'une boule, et semblablement est il des autres estoiles. Por quoy il puet apparoir clerement que la partie de la lune qui est devers le soleil est illuminee, et l'autre non, car la lumiere et les rais du soleil n'i pueent attaindre.

Item: pour ce que la lune est sperique, pres de la moitié de sa superfice est devers nous, est par devers nostre regart, et l'autre partie est aussi comme derriere et non pas vers nostre regart, et ainsi est il de toute chose sperique.

Item: la lune qui par son propre mouvement est meue sous le zodiaque plus tost que le soleil, est aucunefois en conjunction aveques le soleil; c'est quant ele et le soleil sont en .i. meisme degré et en .i. meisme point du zodiaque. Et aucunefois est la lune en opposition; c'est quant elle et le soleil sont en degrés oppositez. Et aucunefois est entre conjunction et opposition et en autres regars. Je di donques que quant la lune est en conjunction ou pres, lors elle est aussi comme entre nous et le soleil. Et adonques la partie d'ele qui est dessus et par devers le soleil est illuminee.

1369 Le ... chapistre] *om.* HKN .xliiii.^{e}] 44/44me MP; 43 D; .xliii. F chapistre] *om.* A croissement] croissent E 1369-1370 Du ... lune] Declaracion plus aplain des choses dessous D[2] 1370 lune] *add.* .xliiii. K; *add.* .xliiii. ca. HN 1371 ce] *om.* DP dit et] *om.* DFGHKLMN 1372 je] *om.* FM declairer] cliarer D comme] commant M enluminee] luminee L; illuminee ABFGKM 1373 mes] *add.* el B est] *om.* M illuminee] enluminee HK 1374 ainsi] aussi DHKN selonc ce] ainsi M; secon ce C ce] *om.* B Averroys] Adverroys L 1375 a] en BHKMNP boule] pole F il] *om.* K 1376 il] *add.* ne P 1377 illuminee] enluminee H car] que D n'i] ne N 1379 Item] Et M pres de] presques M de$_{1}$] que ABCDFGHLMN sa] la FGLM est$_{1}$] *om.* D 1380 est] et ABDFGHKLMN par] *om.* BHN 1380-1381 l'autre ... et$_{2}$] *om.* M 1381 ainsi] aussi D 1383 est] *add.* meue B 1383-1384 ele ... soleil] le soleil et la lune M 1384 .i.$_{1}$] un/une ABDFGHKLMNP degre ... meisme$_{2}$] *om.* FM et$_{1}$] ou GHKN .i.$_{2}$] un/ung ABDGHKNP 1385-1386 Et$_{2}$ ... opposition] *om.* D 1386 en] *om.* D 1387 quant] *om.* D est$_{1}$] *om.* D en] *om.* K ou pres lors] *om.* B elle] *om.* HN aussi] ainsi DM 1388 d'ele] *om.* B illuminee] enluminee HN

1 See Source References," line 1374.

2 Copied from Chap. 45.

Et l'autre partie [qui] est dessous devers nous et devers nostre regart, n'est pas illuminee. Et pour ce, ne po[v]ons nous veoir la lune quant il est conjunction ou pres de conjunction.

Item: [quant] elle est en opposition, lors sommes nous aussi comme entre elle et le soleil. Et adonques est vers nostre regart la partie de la lune qui est devers le soleil et qui est illuminee, et la voions a plain.

Item: quant il y a .i. petit de la partie de la lune qui est devers le soleil, qui est aussi devers nostre regart, adonques appert la lune petite, et tant plus en y a, tant appert elle plus grande. Et appert premierement cornue, et aprés d'autre figure pour
f. 27r cause de ce que elle est sperique, comme dit est.[3] Et selonc ce que dit / Lincoliense,[4] on doit ymaginer .ii. cecles desquels li un devise et separe la partie de la lune qui [est] illuminee d'aveques celle qui n'est pas illuminee. L'autre cecle divise et separe la partie de la lune qui est devers nostre regart d'aveques l'autre partie. Et selonc ce que la lune se approche de la conjunction et se esloigne de l'opposition, selonc ces .ii. cecles participent moins ensemble. Et selonc ce que elle s'aprouce de l'opposition et se esloigne de la conjunction, ces .ii. cecles participent plus ensemble. Et c'est la cause pourquoy la lune appert croistre et ameniser en lumiere. Et selonc ce elle est appellee par divers noms en grec,[5] mais il n'est force de les savoir.

Et la sentence de cest chapistre peut assés apparoir en figure.

1389 partie] *om.* B qui] *om.* ACEFGKLMP dessous] *add.* et BM et$_2$ ... regart] *om.* B regart] *add.* qui AKLMP 1390 illuminee] enluminee H povons] poons CEK nous] *om.* B est] *add.* en BM 1392 quant] *om.* E aussi] *om.* K; ainsi M 1393 est$_1$] *add.* elle P devers] vers M 1393-1394 de ... et$_1$] *om.* B 1393-1394 devers ... est] *om.* D 1394 illuminee] enluminee DN et$_2$] *add.* adoncques nous M 1395 y] *om.* NK .i.] un/ung ABDFHKLMNP partie de la] *om.* FM 1396 tant$_2$] plus HN; *add.* plus B 1397 plus] *om.* BN et$_2$] *add.* puys B 1398 de ce] *om.* BKM ce$_2$] *om.* BF Lincoliense] Licolieuse F; Lincomense HN; Linconite M; Linconiense K 1399 on] en AG; l'en DHKN .ii.] 2 DP; deux BHLMN li un] l'un N; li 1 D separe] *add.* de FGKLMN 1400 est] *om.* ECP illuminee$_1$] enluminee HN d'aveques] d'oncques FHL; de BM d'aveques celle] de l'autre partie de la lune N illuminee$_2$] enluminee HN 1401 de la lune] *om.* B nostre ... partie] nous de celle qui nous estre [*illeg.*] B; *add.* opposite que nous ne pouvons veoir M d'aveques] d'oncques FHL; de MN selonc] secun C ce] *om.* B 1402 la$_2$] *om.* M selonc] secun C; *add.* se GK; *add.* se que D 1402-1403 se$_1$ ... elle] *om.* B 1403 .ii.] 2 DP; deux GHM selonc] secon C s'aprouce ... et] *om.* HKN 1403-1404 moins ...participent] *om.* D 1404 conjunction] *add.* et s'approuche de l'opposition HKN .ii.] 2 P; deux BHLM 1405 c'est] *om.* P la lune] *om.* D selonc] secun C 1406 par] en D 1407 Et ... figure] La figure ensieust B; *add.* s'ens[uite] M; *add.* qui s'sensuit cy dessouz ensuiant N en] *add.* ceste H; *add.* la KM

3 Chap. 44, ll. 1379-1381.

4 See "Source References," line 1398.

5 See the appendix on "Phases of the Moon".

DIAGRAM: CHAPTER 44(A)

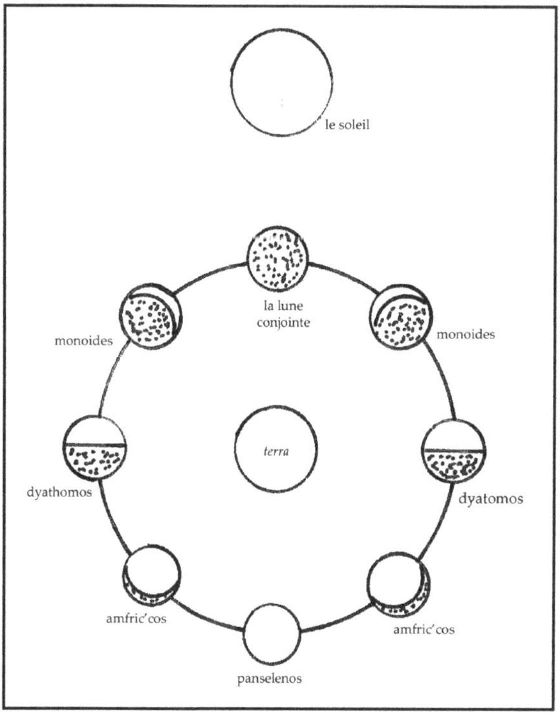

La figure de la lumination de la lune quant a nostre regart

Complete diagram:[6] E CP BFGHKN
No diagram (or space only): DLM

le soleil] *om.* GHKN la lune conjointe] *om.* KN; conjuncion FGH; luna in conjunctione B monoides$_1$] *om.* FGHK; croissant BN monoides$_2$] *om.* BFGHK; decouze N dyathomos$_1$] *om.* FGHKN terra] B dyatomos$_2$] dyathomos P; *om.* BFGHKN amfric'cos$_{1, 2}$] *om.* BFGHKN panselenos] plane lune BFHKN; opposition G

6 Again, the basic diagram is found in mss ECP. A much simpler version is found in the other mss (BFGHKN).

For the derivations of the terms "monoides", "dyathomos", "amfric'cos", and "panselenos", see the appendix on "Phases of the Moon".

DIAGRAM: CHAPTER 44(B)

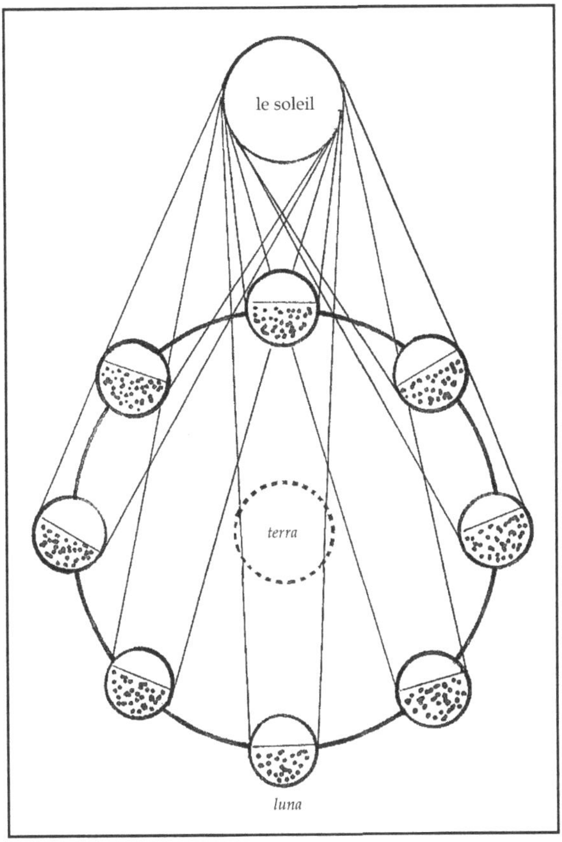

La figure comment la lune est illuminee du soleil

Complete diagram: E CPM
No diagram (or space only): BDFGHKLN

le soleil] sol M terra] M luna] M

Le .xlve chapistre. [De la] declaracion plus plaine des choses dessous dites

Toutesfois que .i. corps sperique plus grant illumine un autre corps plus petit, il enlumine plus que la moitié. Et aussi tant sont cil .ii. corps sperique plus pres l'un
f. 27^{v} de / l'autre, tant plus grant partie est illuminee de cil corps plus petit. Et tout ce est demonstré et prouvé en la science de perspective.

Or est il ainsi que le soleil est merveilleusement plus grant que la lune, et sont .ii. corps speriques, par quoy il s'ensuit que le soleil illumine plus que la moitié de la lune.

Item: pour ce que la lune est plus pres du soleil en temps de conjunction, il s'ensuit que adonques en est plus grant partie illuminee que quant elle est en opposition, ne que en autre fois.

Item: aussi s'ensuit il que tant plus de la lune nous appert illuminee, et moins en est illuminee, et tant moins tant plus.

Item: selonc perspective, de chascune chose sperique qui est plus grande que l'espace d'entre nos .ii. oiels, l'en ne puet veoir fors moinz que la moitié, car de cele chose, moins que la moitié en est devers nostre regart. Par quoy il s'ensuit que, posé que la lune fust illuminee toute environ, si ne pourrons veoir fors moins que la

1408 Le ... chapistre] *om.* HKN .xlve] 45/45^{m} MP; 44 D De la] *om.* ECFGL; De PA plus] *om.* BFGLM plaine] plainniere GL 1408-1409 Le ... dites] *rep.* C De ... dites] Des causes des eclipses en general D[1] 1409 dites] *add.* .xlv. K; *add.* .xlv. cap. HN 1410 .i.] un/ung ABDFGHKLMNP illumine] enlumine PN; enlumiere H autre] *om.* D corps$_2$] *add.* sperique ABCDFGHKLMN il] *om.* HN 1411 enlumine] est illumine B que] *rep.* K; de B aussi] ainsi FM cil] ycieulx B .ii.] 2 D; deux PBM 1412 plus$_1$] *om.* D illuminee] enluminee HKN(*and so elsewhere*) de cil] du M plus$_2$] *om.* HKMN; qui est plus B Et tout] *om.* B 1413 prouve] demonstre M science de perspective] seconde partie de sperique H; seconde partie de sperique et perpective N 1415 .ii.] 2 DP; deux B que$_2$] de M 1417 soleil] *om.* B conjunction] *add.* du soleil que en autre temps B 1418 en$_1$] que la lune B grant] *om.* B 1418-1419 quant ... fois] en temps de opposition ou d'autre temps B 1419 que en] *om.* PL; que ACFGKM 1420 et] tant B 1421 est illuminee et] *interlin.* D; y a et B tant moins] *om.* D 1422 selonc] secun C perspective] spectulative H; par speculative N que] de que B 1423 .ii.] deux BHKMP; y D ne] *interlin.* G que] *om.* K; de M cele] telle HLN 1423-1424 car ... regart] *om.* B 1424 que] de DKM en] *om.* L Par] Pour D 1425 que$_1$] *om.* D toute] *add.* enluminee tout EP C(*and erased*) environ] asenviron M; entour N ne] n'en AMN pourrons] *add.* nous ABCDGHKLMN moins] *om.* B; *marg.* G que$_2$] de M

1 Copied from Chap. 46.

moitié. Et il est ainsi que plus de la moitié d'elle est enluminee, comme dit est. Et pour ce, est ce impossible que nous voions de la lune autant comme d'elle est illuminee. Et, se nous appelons "plaine lune" quant tout ce qui est vers nostre regart et qui nous puet apparoir est illuminé, il puet bien estre plaine lune nonobstant que il ne soit pas opposicion.

Mes il couvient que ce soit assés pres de vraie opposicion. Et adonques tout ce que nous voions de la lune et po[v]ons voier est partie de ce que en est illuminé. Et tout ce pourroit estre legierement demonstré par perspective.

Or appert il donques par ce qui est dit comme la lune est illuminee et comme elle ne croist pas ne n'appetice en la maniere que elle appert croistre et appeticier.

Et des autres estoiles ou planetes, l'en doit savoir que ja soit ce que elles soient enluminees du soleil qui est fontaine de lumiere, toutevoies ne apparoissent il pas croistre et appeticier en lumiere par elles approchier ou esloignier du soleil, comme fait la lune. Et la cause est car il sont par dessus et plus haut que le soleil. Et ainsi la partie d'elles qui est devers le soleil et illuminee est par devers nous ou par devers nostre regart. Et aucunes planetes sont dessous et plus bas que le soleil, si comme sont Venus et Mercure selonc plusieurs. Toutevoies ne apparoissent il pas croistre
f. 28r et appeticier en lumiere comme la lune, ou pour / ce que il apparoissent petites ou

1426 Et ... moitie$_2$] *om.* M il] *om.* P de] que ACDFGHKN est$_2$] *om.* M enluminee] illuminee DGKLMN 1427 ce$_2$] il BD d'elle] el B; delle D; *add.* en K 1428 illuminee] *add.* comme dit est D Et] Item ADFGHKLMN Et se] Item B 1768 lune] *om.* DM tout] *om.* B 1429 et] ou L lune] *om.* L 1429-1436 il ...ce] *illeg.* B[2] 1430 pas] *om.* D 1432 povons] poons E que$_2$] qui DF illumine] enlumine DL 1433 par] *interlin.* D perspective] par spective DN 1434 qui] que F 1436 que ja soit ce] *om.* D elles soient] aussi comme la lune est B 1437 enluminees] illuminees BGKLM il] ne D L 1437-1441 toutesvoies ... regart] aussi pour toute les autres planettes. Et toutefois il n'acroissent pas ou n'appetissent pas acroistre en lumiere ou appetissent aussi *illeg.* la lumiere et la *illeg.* car celles qui sont au desus du soleil nous apparissent tousjours illuminee *illeg.* ainsi sont que le soleil est entre nous et elles. B 1438 et] ne DL; ou N appeticier] *illeg.* D elle] ele E ou] ne N comme] *add.* est 1439 car il] pour ce qu'il D il] elles HKN par] *om.* L et$_1$] *om.* KL ainsi] aussi D 1439-1440 ainsi ... et] *om.* M 1440 et] est L illuminee] *add.* qui L par$_2$] *om.* F 1441 Et] *add.* se CFGHKLN aucunes] les B; si autrement M sont] qui sont au B dessous] audessoubz N et$_2$] ou M et$_2$ ... si] aussi B 1442 selonc] secou/secon C selonc ... Toutevoies] *om.* B il] *om.* BM 1443 et] *illeg.* D; ne HKMN; ou ABFGL lumiere] *add.* aussi B lune] *add.* pour aucune de plusieurs cause B que il] qu'elles HKN

2 B deviates most (and uniquely) from the base text, as well as being written in a sloppy 15th-century hand. These 10 lines in the ms meander back and forth over the standard content with many words so abbreviated or so badly written that their sense is impossible to decipher. The rest of the chapter (lines 1436-1451) also has these problems but is closer to the base text.

regart de la lune, ou pour ce que il ne s'esloignent gueres du soleil, ou pour ce que la lumiere du soleil entre dedens elles. Et ne sont pas corps opacques ou obscurs tant comme est la lune, mes sont corps transparens aussi comme cristal si que la lumiere du soleil entre dedens eus parfont aussi comme en une lampe plaine de oile clere, et ne sont pas tant seulement illuminees le dehors ou vers la superfice aussi comme est la lune. Et pour ce ne apparoissent pas de figure cornue ne croistre ne appeticier en [lumiere], comme dit est. Et par aventure reluisent il, et ont [lumiere] d'elles meisme.

1444 ce] *om.* D   que il] qu'elles HKN   gueres] gueuc queres A   gueres du soleil] par comme la lune B   1445 elles] *om.* BN   corps ... ou] *om.* B   tant] *om.* B   1446 est] *om.* P   aussi] *om.* B; ainsi M   si] aussi B   lumiere] clarte B   1447 eus] en DFGHKL   parfont] apparaissent B   aussi] ainsi M   plaine] clere B   1448 illuminees] *om.* D; enluminees HN; *add.* vers BDFGHKLMN   aussi] ainsi M   1449 ce] tant B   apparoissent] *add.* il/ilz FLM   de ... ne$_2$] *om.* B   ne$_3$] ou ABFGHKMN   1450 lumiere] lumine EC   comme dit est] *om.* B   reluisent il et] qu'il B   lumiere] enlumine EC   d'elles] de eulx BM

LE .XLVI.E CHAPISTRE. DES CAUSES DES ECLIPSES EN GENERAL

Chascun qui ha raison en soy puet veoir clerement que, puisque la terre est ou centre du zodiquae et ou centre de la ligne ecliptique, dont j'ai parlé ou .x.e chapistre,[1] il s'ensuit que se la lune et le soleil estoient meus sous celle ligne ecliptique il couvendroit en chascune conjunction de lune et de soleil que la lune fust entre la terre et le soleil. Et ainsi seroit il en chascune conjunction eclipse de soleil, car eclipse de soleil est causee de ce que la lune est entre la terre et le soleil, ou entre nous et le soleil.

Item: semblablement couvendroit il que en chascune opposition la terre fust entre la lune et le soleil. Et c'est cause de eclipse de lune. Et aussi s'ensuiroit il que il fust en chascune opposition eclipse de lune, et il n'est pas ainsi. Et pour ce, doit l'en savoir aussi comme il fu dit au .x.e chapistre,[2] le soleil ha son mouvement parmi le milieu du zodiaque sous la ligne ecliptique, et les autres planetes ne sont pas meues sous celle ligne, mes declinent et se traient plus vers septentrion d'une part, et plus vers midi d'autre, selonc la latitude du zodiaque. Et par especial, je vueil dire quant a ce de la lune car elle ha son propre mouvement selonc une ligne laquelle interseque et devise l'ecliptique en .ii. poins opposites; et le point ou la lune
f. 28v commence a decliner hors d'ecliptique / vers septentrion est appellé "Caput Drachonis"; et l'autre ou elle se prent a decliner hors l'ecliptique vers midi est

1452 Le ... chapistre] *om.* HNK .xlvl.e] 46/46me MP; 45e D chapistre] *om.* A Des ... general] De l'eclipse du so[leil] en general D[3] Des causes] *om.* H general] *add.* .xlvi. K; *add.* .xlvi. ca. HN 1453 est] *om.* K 1454 .x.e] 10/10me DMP 1455 se] *om.* GL 1456 en] que en B de lune] *om.* G 1457 soleil] *add.* et entre nous et le soleil M ainsi] adoncques B; aussi D il ... conjunction] *om.* B 1458 de ce que] par la interposition de B est$_2$] *om.* B 1458-1459 ou ... soleil$_2$] *om.* B 1460 convendroit] *rep.* L; conmendroit H opposition] *add.* que DFGKLMN 1460-1462 il ... opposition] toutefois que nousavon la lune en opposition que la terre soit entre le soleil et la lune adonques soit B 1461 entre] om. G aussi] ainsi AMG il] *om.* HKN 1463 aussi] que B; que si ADFGHKN .x.e] 10/10me DMP son] *add.* propre B parmi] par B 1464 la ligne] le signe N 1466 d'autre] *add.* part N selonc] secun C 1467 car elle] qu'elle DHN selonc] secon/secou C laquelle] qui B 1468 .ii.] 2 DP; deux BFHMN 1469 d'ecliptique] l'ecliptique ABFGHKLMN; de l'ecliptique D Caput] Cauda B 1469-1470 septentrion ... vers] *om.* GL 1470 Drachonis] Draconis HKN l'autre] *add.* poins/point BFN se] *om.* M prent] commence ADFM a] *om.* D hors] *add.* de D

---

1 Chap. 10, ll. 302-319.
2 Chap. 10, ll. 306-307.
3 Copied from Chap. 47.

appellé "Cauda Drachonis".[4] Et puet estre monstré en figure. Et par ces choses appert que quant il est conjunction, et la lune ne est en aucun de ces .ii. poins, combien que le soleil et la lune soient en .i. meisme degré ou minut du zodiaque, et en .i. meisme meridian, toutesvoies la lune, selonc la latitude, est plus vers septentrion que le soleil, ou plus vers midi. Et ainsi n'est elle pas entre le soleil et la terre. Et pour ce n'est il pas eclipsé de soleil. Et semblablement quant il est opposition et elle n'est en .i. de ces .ii. poins dessus dis ou pres, lors n'est pas la terre entre la lune et le soleil et pour ce n'est il pas adonques eclipsé de lune.

1471 Cauda] Caput B Drachonis] Draconis HKN Et$_1$] *add.* ne D Et$_1$ ... figure] *om.* M; monstré il apres en la figure B en] *add.* ceste HKN choses] *add.* il B 1472 que quant] il quant ADFGHKMN; il quant que L est$_1$] puet estre GL de ces] des K; de ses DN .ii.] 2 D; deux BHMNP 1473 .i.] un/ung ABDFHKLMNP du] ou du N 1473-1474 e t$_2$ ... meridian] *om.* B 1475 .i.] un/ung ADFGHKLMNP la$_2$] *om.* ADFL vers] devers GL 1475 plus ... midi] mains B 1476 quant] *rep.* P 1476-1478 Et$_2$ ... lune$_2$] en chascune coniunction, ne eclipse de lune en chascune opposition pour ce que la lune ne se tienne pas en aucun de ces deux pointz dessus dits ou pres d'iceulx M 1477 .i.] un/ung FDHKNP; aucun B ces] ses N .ii.] 2 D; deux BHP ou pres] *om.* N 1478 entre] *add.* la terre B et$_2$ ... lune$_2$] *om.* F

4 The orbit of the moon is tilted about 5° from the orbit of the sun (the ecliptic) and the two cross twice a year at the "lunar nodes." The ascending node (called the "Caput Draconis" / Dragon's Head) is when the moon moves to the north of the ecliptic; the descending node (the "Cauda Draconis" / Dragon's Tail) is when the moon moves to the south of the ecliptic. Lunar eclipses can occur only when the moon is at or near one of its nodes.

Because the orbit of the moon precesses around the ecliptic, the time of the nodes changes from year to year. The cycle of precession is completed approximately every 18.6 years.

DIAGRAM: CHAPTER 46

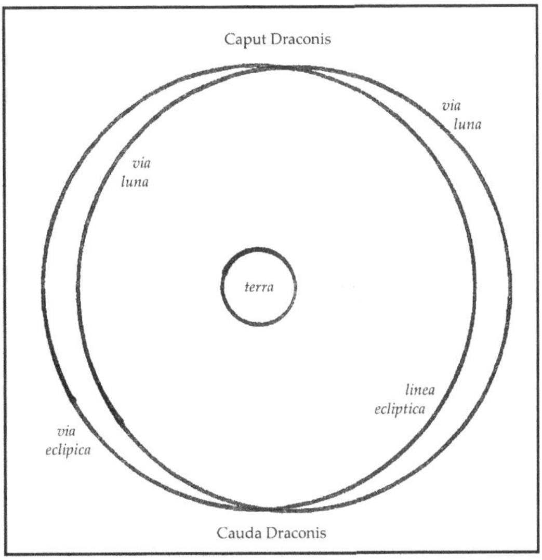

Complete diagram:[5] ECP BFGHKMN
No diagram (or space only): DL

via lune$_{1,2}$] M terra] M via ecliptice] M; Centrum Drachonis B linea ecliptica] M Cauda Draconis] Cauda F

5 Except for ms M, the diagrams in the mss simply show two intersecting circles with "Caput Draconis" and "Cauda Draconis" marked on them. This minimalist presentation tries to represent the intersection of the two "orbits" of the sun and the moon, without indicating which orbit belongs to which, nor what is "north" and what is "south," nor in which directions the sun and the moon are moving.

In ms M, some attempt has been made to define the elements of the diagram, with labels on the two orbits and the positioning of the earth in the middle. In general, diagrams in ms M have more (or different) information than the ECP group or the BFGHKN group.

LE .XLVIIE CHAPISTRE. DE ECLIPSE DE SOLEIL EN ESPECIAL

La lune, quant elle est de soy, est .i. corps obscur, et est illuminee du soleil en sa superfice, si comme dit est au .xliiii.^{e} chapistre,[1] par quoy il couvient que elle porte ombre derriere soy. Et pour ce que la lune est mendre que le soleil, celle umbre procede loing de la lune, et va tousjours en appetissent a la maniere du pignon d'un f. 29^{r} clochier d'eglise. Et ne se estent pas celle ombre ne ne va pas si/ques ou ciel de l'autre part a l'opposite. Et pour ce, ne puet celle umbre touchier ne attaindre siques as estoiles, ne les eclipser. Mes celle umbre s'estent bien aussi loing ou plus, comme la terre est loing de la lune.

Et pour ce, quant il avient que, a heure de la conjunction, la lune est en un de ces .ii. poins dessus dis, qui sont appellés "Caput" et "Cauda Drachonis", ou pres, lors son umbre touche et couvre aucune partie de la terre, et adonques est il eclipsé de soleil en ceste partie de la terre. Et ainsi comme l'ombre de la lune couvre partie de la terre ou temps de l'eclipse du soleil.

Aussi adonques la lune muce le soleil en tout ou en partie, et le couvre ou regart de nostre voiement. Et est asavoir que il ne couvient pas, quant il est eclipsé de soleil en .i. lieu, que eclipse soit partout, ne que elle soit aussi grande une fois

1479 Le ... chapistre] *om.* HKN .xlviie] 47/47me MP; 46 D De ... especial] De [*illeg.*] de la lune D eclipse] l'eclipse BFGHLMNP especial] *add.* .xlvii. K; *add.* .xlvii. chap. HN 1480 elle] *del.* C; *om.* ABDFGHKLMN .i.] un/ung ABDFHKLNP 1481 si] *om.* DGHKLN si ... au] comme il appert au B .xliiii.^{e}] 44/44me DMP; .xl. B par] pour KN 1482 lune] *add.* ne F que$_{2}$] *add. and del.* la lune C 1483 appetissent] *add.* l'ombre de la lune B a] en DHLKMN maniere] mesure D du] d'un AKLM du pignon] *om.* HN 1484 clocher d'eglise] eglise B estent pas] soit D pas$_{2}$] *om.* D siques] jusques DHKLMNP 1485 l'opposite] *add.* en la huitieme espere M siques] jusques BDHKLMNP 1486 eclipser] *add.* c'est a dire cause/causer en elles/icelles eclipse ABDGHKLMN 1488 un] aucun B 1489.ii.] 2 D; deux BHMP dessus dit] *om.* D Drachonis] Draconis HKN pres] *add.* d'ilec M 1490 couvre] cueuvre N(*and elsewhere*) 1490-1491 et$_{2}$ ... terre] *om.* BP 1491 ainsi] aussi ADFGK 1493 Aussi ... soleil] *marg.* N(*later hand*) muce] *add.* et cache M tout] tant L en$_{2}$] *om.* M 1494 voiement] veoir M 1495 .i.] un/ung ABDFHKLMNP lieu] *add.* que il ne convient pas D partout] sus toute la terre M ne] je C grande] *add.* a KN

1 Chap. 44, ll. 1371-1378.

comme autre, ne en une meisme fois que elle soit grande en .i. lieu comme en autre. Car en aucun lieu et aucune fois, l'ombre de la lune couvre et muce, en est entre une partie de soleil et nous, et autre fois ou en autre lieu en couvre plus grande ou plus petite partie, selonc ce que la lune est plus ou mains pres d'un de ces .ii. poins dessus f. 29v diz, / et selonc divers regars que la lune ha diverses habitacions ou a divers lieux. Et ceste diversité de regars puet estre sceus parla theorique des planetes, et par les tables d'astrologie theorique speculative des mouvemens des cecles et des regars, des autres accidens du ciel.

1496 comme$_1$] que M; *add.* a GHKN; *add. interlin.* a A autre$_1$] l'autre P ne ... autre$_2$] *om.* B elle] eclipse D soit] *add.* aussi ADFGHKLMN .i.] un/ung ADFHKLMNP comme$_2$] *interlin.* D en$_3$] *add.* ung M autre$_2$] l'autre DP 1497 en$_1$] *om.* N aucune] autre N en$_2$] ou ABDKLM en est] que F entre] autandre B 1498 de] du ABDHKL de soleil] *om.* FM ou$_1$] *om.* M en$_1$] *add.* un DHN grande] *add.* partie L 1498-1499 plus$_1$ ... partie] au plus ou mains B 1499 selonc] secun C est] *add.* ou F ou mains] *om.* B d'un] *om.* ABDGHKLN; de F ces] ses N .ii.] 2 DP; deux BMN 1500 diz] *interlin.* A selonc] secun C que la lune ha] et M ha] *add.* a CGHKLN a] *om.* BFMN lieux] *add.* que la lune peult avoir M 1501 sceus] de clairer D et] ou H 1502 theorique] Theorique est la science ADFGHKLMN

DIAGRAM: CHAPTER 47

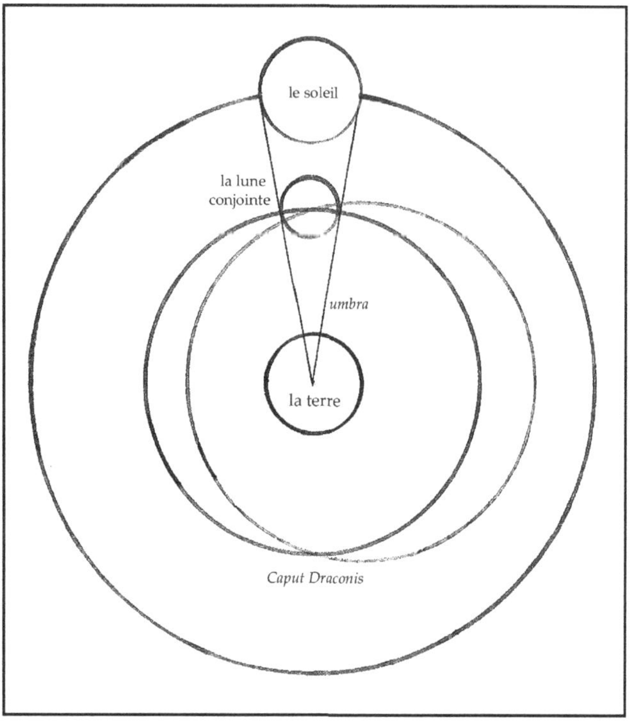

La figure de l'eclipse du soleil

Complete diagram: E CP BDFGHKMN
No diagram (or space only): L

le soleil] sol BDKM   la lune conjointe] lune DK; luna BGM; la lune HN   umbra] M[2]   la terre] terre DK; terra BGM   Caput Draconis M   La figure ... soleil] *om.* BDFGHMN; Eclipse de soleil K

2 In ms N, the moon is large enough to "eclipse" the entire earth, and the word "umbra" is set in the extension of the moon's shadow below the earth.

LE .XLVIIIE CHAPISTRE. DE L'ECLIPSE DE LA LUNE EN ESPECIAL

Pour ce que la terre est moult plus petite que le soleil, son umbre procede derriere soy et va en appetissant et en agreillissant, si comme il fu dit de l'ombre de la lune ou chapistre precedent.[1] Et est tousjours la pointe et le bout de celle umbre a l'opposite du soleil est appellé "nadair solis". Et se extent l'umbre de la terre sique a l'espere de la lune. Et donques toutesfois que la lune chiet en l'umbre de la terre et en l'eclipse, il couvient que elle soit sous l'ecliptique ou pres. Et donques couvient il qu'elle soit en .i. des .ii. poins dessus dis ou pres. Et pour ce, avient il que aucunefois partie de la lune est eclipsee, une fois petite, et autre grande; unez fois d'une part, et autre d'autre. Et aucunefois est toute eclipsee et dure l'eclipse une fois plus, autre moinz, selonc ce que elle entre ou chiet en l'umbre de la terre d'une part, ou d'autre, toute ou partie, et plus ou moins. Et puet apparoir a figure. Item: on puet veoir clerement par les choses dessus dictes que, quant la lune est eclipsee, elle ha deffaute de lumiere en soy meismes, et est eclipsé d'une quantité et d'une maniere f. 30r par tous pais, aussi en .i. lieu / comme en autre, ja soit ce que l'en ne la voie pas par touz pais et que l'en la voit a une heure en .i. lieu, et a autre en autre, et en .i. meismes moment de temps. Quar quant il est soleil couchant en .i. lieu, est une nuit

1504 Le ... chapistre] *om.* HKN .xlviii.^{e}] 48/48me MP; 47^{e} D De l'eclipse] *om.* F De$_1$ ... especial] De la lune et du soleil et de l'ombre de la terre D la] *om.* K especial] *add.* .xlviii. K; *add.* .xlviii. ca. HN 1505 moult] *om.* D 1506 va] le ha B de$_1$] en D 1507 tousjours] *om.* BD 1508 soleil] *add.* et cil point opposite du soleil BCFGHKLN; et cellui point opposite M sique] jusques DHKLMNP 1509 donques] *om.* M; adonc D 1510 l'eclipse] eclipse AFHLM soit] *om.* G pres] *add.* d'ilec M 1511 il$_1$] *interlin.* A .i.] un ABDFHKLMNP .ii.] *om.* K; 2 D; deux BHMNP pour] *om.* D il$_2$] *om.* D 1512 partie de] *om.* D une] aucune F et autre grande] autre fois grande ABDFGHKLMN 1513 autre$_1$] *add.* foiz DFGKLMN aucunefois] aucune C est] *add.* elle ADFGHLN 1513-1514 toute ... moinz] ce plus eclipse et dure plus a une fois la eclipse que a l'autre B 1514 autre] l'autre K; *add.* foiz ADLM selonc] secun C ce] *om.* B entre ou] *om.* M ou chief] *om.* B 1515 et$_1$] *om.* K; ou BDHN ou$_3$] et ACFGLM et$_2$] si comme M a] en GHKLN; en ceste M Item] *om.* M on] l'en B; l'on M 1517 meismes] *om.* B eclipse] l'eclipse CKL 1518 aussi] *add.* bien P .i.] un/ung ABDFHKLMNP l'en] l'on M; ne HN; en/on D la] *om.* CP 1519 que l'en] que l'an G; que *corr. to* qu'on D en .i.$_1$] a une L; en un/ung ABDFHKMNP autre en autre] autre heure en autre heure B en$_2$] oung P en$_3$] *om.* D .i.$_2$] *om.* BGHKLN; ung/un DFP 1519-1520 autre$_1$ ... meismes] autre heure en ung autre et en un autre M 1520 Quar] Car HN .i.] un/ung ABFHKLMNP est une nuit] il est nuit ABFGKLMP; il est une nuit HN; est minuit K; il n'est une nuit D

1 Chap. 47, ll. 1480-1487.

en une autre lieu plus vers orient. Mes, a revenir a propos, le soleil pour son eclipse n'a point en soy defaute de lumiere, car sa clarté dure touzjours en perpetuité sans faillir. Et pour ce, quant nous disons que il est eclipsé de soleil, nous deusson mieux dire, a parler proprement, que il est eclipsé de terre. Car aussi comme la terre en eclipse de lune fait umbre a lune, tout aussi la lune en eclipse de soleil fait umbre a la terre.[2]

Fors tant que la terre porte empeechement de lumiere a la lune plus souvent que la lune ne fait a la terre en certaine pre, comme a Paris ou a Romme. Et avient souvent que l'umbre de la terre couvre toute la lune, mes l'umbre de la lune ne puet onques couvrir toute la terre.

1521 lieu] *om.* B lieu ... orient] *om.* N Mes a] *om.* N a_1] *om.* K; ou D a_2] *add.* mon HN pour] par B eclipse] *add.* en lieu B 1521-1522 pour ... car] na dessus point de lumiere par soy que D 1522 en soy] *om.* M; de B; *add.* de N en_2] a DL 1523 que ... $nous_2$] *om.* F il ... soleil] le soleil est eclipse M 1524 il] *om.* B de terre] *om.* M aussi] ainsi M 1525 de lune] *om.* K aussi] ainsi M 1527 tant] *om.* D 1528 pre] partie ABDFGHKLMN; *add.* si F 1529 $lune_2$] terre D; *corr. from* terre L 1530 onques] bonnement D; jamais M toute] *om.* N

2 In an eclipse of the moon, the earth blocks the light of the sun; so when the moon blocks the light of the sun, it should be an "eclipse of the earth". But we talk of these events, not from the point of view of somewhere out in the universe, but from our point of view here on earth – hence "eclipse of the sun."

DIAGRAM: CHAPTER 48

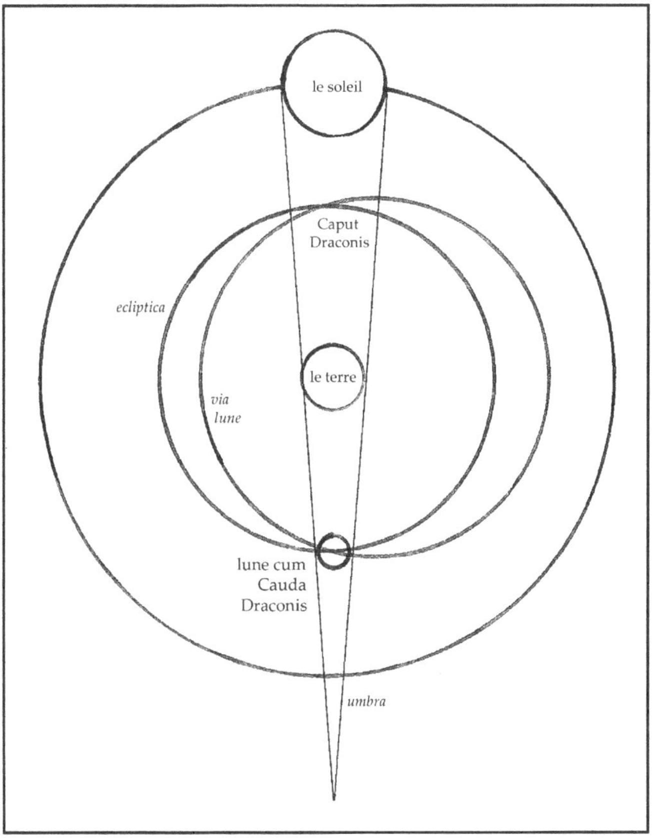

La figure de la lune eclipsé

Complete diagram: E CP BDFGHKMN
No diagram (or space only): L

le soleil] sol BDKM Caput Draconis] *om.* BDFGHKN; cauda M ecliptica] M la terre] terra BDM; terre FK via lune] M lune cum Cauda Draconis] *om.* B; luna D; luna caput M; lune FHK; la lune G umbra] N La figure ... eclipse] *om.* BDFGHMN; Eclipse de lune K

LE .XLIX.[E] CHAPISTRE. DES COULEURS DE LA LUNE EN TEMPS D'ECLIPSE, ET DE L'ECLIPSE DES AUTRES ESTOILES

La lune quant elle est eclipsé, retient en soy aucume couleur ou lumiere. Et la cause est pour ce que le soleil illumine l'espere des .iiii. elemens qui sont de matere espesse ou dempse, pour laquelle chose l'espere du feu et l'espere de l'air recoivent lumiere du soleil et la retornent ou reflectent en l'umbre de la terre et sus la lune ou temps de l'eclipse, et ainsi les autres planetes et estoiles rayent sus la lune quant el est eclipsé. Et semblablement est la terre illuminee de nuit sans la lune pour les causes dessus dites, fors quant il y a empeechement de nues ou de obscurtés de vapeurs.

Item: la lune en temps de eclipse appert une fois d'une couleur, autre fois d'autre. Et je cuide que la cause soit pour la lumiere que elle recoit des planetes et des estoiles, et de la reflexion de l'air, comme dit est. Laquelle lumiere est une fois d'une maniere, et autre fois d'autre, selonc divers regars que la lune ha aus autres estoiles en divers temps, et selonc la diversification de la re[f]lexion de lumiere causee de l'air ou d'aucunes vapeurs. Et selonc ce, est elle une fois d'une couleur, et autre fois d'autre, et aussi mue elle couleur selonc ce que elle est plus ou mains avant en l'umbre de la terre.

Et pour ce, quant a jugier de la signification de l'eclipse de la lune, regardent les astrologiens as couleurs de la lune ou temps de l'eclipse. Item: eclipse, a parler
f. 30v proprement, est defaute de lumiere / et ainsi est aucunefois la lune eclipsé, et aussi

1531 Le ... chapistre] *om.* HKN .xlix.[e]] 49/49me MP; 48 D 1531-1532 Le ... estoiles] *marg.* C; Des eclipses et des coleurs 2 [*illeg.*] le 48 chapitre de ses eclipses et de ses couleurs 2 [*illeg.*] D 1532 en ... estoiles] *om.* N et ... estoiles] *om.* FM; *add.* .xlix. K; *add.* .xlix. chap. HN autres] *om.* H 1533 ou] en HKN 1534 illumine] enlumine DHN .iiii.] 4 DP; quatre BHM 1535 dempse] condempse L l'espere$_2$] *om.* D 1537 de l'eclipse] d'eclipse DLMN ainsi] aussi ABDFGHKMN rayent] ont rayeur H sus la lune] *rep.* N 1538 est$_1$] *add.* en L illuminee] *om.* B; enluminee N la$_2$] *om.* N 1539 fors] *erased* B de obscurites] du elevation B 1541 de eclipse] d'eclipse ADFGKLMNP; de l'eclipse BH 1543 de$_1$] *om.* M reflexion] inflexion F est$_1$] *om.* K 1544 selonc] secun C 1545 estoiles] autres planettes K selonc] secon/secou C reflexion] replexion CE de$_2$] *add.* la D 1546 causee] comme N selonc] secon C ce] *om.* G une$_1$] de une G 1547 d'autre] autre H aussi] ainsi DFP selonc] secun C que elle] qu'il D est] *add.* ou F 1549 ce] *om.* F 1550 astrologiens] astrolo//astrologiens N as] es K; les GL de l'eclipse] d'eclipse GL 1551 ainsi] *om.* B la lune] *om.* D aussi] ainsi K

la terre en partie eclipsé, quant nous dison que il est eclipsé de soleil. Et nule autre planete ou e[stoi]lle ne puet estre eclipsé, comme dit est ou .xlve chapistre.[1] Mes eclipse puet estre p[ri]se autrement ou regart de nous. Et ainsi une chose est eclipsé quant une autre chose est entre nous et elle qui empeeche que nous ne la voion, et la fait eclipser quant a nous. Et en ceste maniere puet la lune eclipser nous le soleil. Et aussi nous eclipse elle aucunefois une autre planete, ou une estoile ou plusieurs. Et en ceste maniere une planete puet eclipsier une autre planete ou estoile qui est par dessus soy et en plus haut ciel. Mes ce n'est pas proprement eclipse.

1552 eclipse$_1$] *om.* HN 1552-1553 Et ... eclipse] *om.* P 1553 ou$_1$] ne B estoille] elle EC ne] *om.* D .xlv.^{e}] .xlix. (?) E[2]; 45/45me DMP; .xlvi.^{e} H 1554 prise] puse E; p'us B; *illeg.* H ainsi] aussi D 1554-1555 eclipse ... est] *om.* K 1556 puet] fait B; nous puet K eclipser] *add.* a HN nous$_2$] *add.* et DHN 1556-1557 peut ... Et] la lune nous peult eclipser M nous ... plusieurs] le soleil et aucunes estoile B 1557 aucunefois] autrefois N 1557-1558 ou$_2$ ... estoile] *om.* D 1557-1559 eclipse ... ciel] peult elle eclipser les autres planetes et les estoiles estans au dessus et a l'endroit d'elle M 1558 une$_2$ ... estoile] aultre B 1558-1559 par ... et] *om.* B

1 Chap. 45, ll. 1436-1445.

2 The chapter number is unclear in ms E – it could be ".xlix." or ".xlv." The final dot is raised so that it might make the final "v", i.e., "u" (with its serif at the bottom right and the dot at the top right) look like "ix".

.L.^{E} CHAPISTRE. LA FIN DE CESTE OEUVRE

Je vueil ici faire fin, car je ne vueil ici parler des epiceles ne des excentriques des planetes, ne des autres fortes choses, pour ce que ce seroit trop longue chose, et que seroit pas aesié a traitier en francois souffisaument. Et il me semble que il vault miex taire s'en que parler en sans monstrer les causes et les neccessités pourquoy telles choses furent trouvees, et la possibilité ou la maniere comment il pueent estre.

Item: mon propos n'estoit pas d'entrer plus avant en teles su[b]tilités, ne de baillier astrologie en francois, mais tant seulement declairer grossement la disposition en general de l'espere du monde. Et ce que de ce est honneste assavoir a tout homme, et par especial a prince de noble engin, fors tant que il n'en lesse nullement a faire chose qui appartiengne a son office ou estat, et au gouvernement de la chose publique.

Et se il se vouloit profunder plus avant quant a la speculative des jugemens, ce seroit curiosité quant a lui, et chose ou il ne doit point mettre son entente. Et se il en vouloit affectueusement savoir et enquerir quant a la practique des jugemens des fortunes a venir, ce seroit chose nient certaine, inpertinente a lui, et perilleuse quant a Dieu et au monde et se mettroit en peril de perdre ame et corps, et biens et honneur

1560 .l.^{e}] La .l.^{e} AG; Le 50me P; Le cinquantiesme M; Le 49^{e} D .l.^{e} chapistre] *om.* HKN La] de la BHN; est la GL; qui traite de la M La ... oeuvre] De la fin de ceste et causes dessous dit D ceste oeuvre] ceste presente oeuvre LG; d'euvre .l. chap. H; de l'euvre. .l.^{e} c. N; *add.* .l. K 1561 vueil$_2$] *add.* pas ADFGHKLM epiceles] epicicles ADFGHLMN 1561-1562 des$_1$ ... ne] *om* B epiceles ... autres] *om.* K 1562 planetes] palenes H fortes] *om.* D ce] *om.* H seroit] soit H longue] forte B que] *add.* ne ABFGKLMNP 1563 seroit pas] ne soit D pas] *om.* N souffisaument] *om.* B Et] Et aussi B Et ... semble] *rep.* B 1564 s'en] en soy D parler en] d'en parler M telles] celles B 1565 comment] comme ABFH comment il] comme telles choses L pueent] puet B; peult M 1566 Item] Aussi M teles] celles B subtilites] supertilites ECP 1567 declairer] de declairer N 1568 de l'espere] *om.* B que de ce] de quoy L 1569 homme] *add.* du monde B par] *om.* H tant] *add.* seulement ADFGLM n'en] na en M lesse] lespere M 1570 ou] *add.* a son HN et] ou L 1572 se il] cil N quant] en H des] de ces M jugemens] mouvemens DFGHKLMN; *corr. to* mouvement A 1573 point] pas ABFLMN; *add. and del.* avoir C en] *om.* B 1575 fortunes] *add. illeg.* B a venir] *om.* M nient certaine] incertaine M 1576 a Dieu] a lui/a liu GL peril] danger M

– si comme j'ai plus a plain declairé et prouvé en .i. livret[3] en francois que j'ai fait a cest propos et sus ceste matiere.

Item: j'ai parlé en cest traitié, en aucuns liex, prolixement et ay esté long afin que chascun de bon entendement puisse ce que j'ai dit legierement entendre et sans expositeur. Et encor pour ceste cause ai je ici en la fin faite une table des mos f. 31r estranges qui sont en cest traitié, en laquelle table je signe / les chapistres ou tiels mos sont exposés et les met selonc l'ordre de l'A, B, C, affin que, quant l'en treuve .i. tel mot en aucun chapistre, l'en puisse avoir recours et trouver a aisiement le chapistre ou quel tel mot est devant exposé. Car chascun est exposé ou diffini ou chapistre la ou il est premierement trouvé.

1577 j'ai] *add.* dit D   declaire] *rep.* K   et prouve] *om.* B   .i.] un/ung ABFHKLMNP   1579 en cest traictie] *om.* D   liex] *om.* K; *add.* moult GL   1580 legierement] *om.* B; sigierement N   1581 encor] aussi BD   1582 je] j'ay BL   tiels] tieux N   1583 selonc] secun C   l'A] A H   l'A, B, C] l'abece B; *add.* D D   1584 .i.] un/ung ABDFHKLMNP   a] *om.* ABDGHKLM   1585 ou quel tel] en qu'il le N   tel] le BH   devant] *om.* M   expose$_1$] *add.* ou diffini HKN   Car] que D   chascun] *add.* mot ADFGHKLM   1586 la] *om.* GL   trouve] *add.* Explicit HMN; *add.* Explicit liber F; *add.* Et hoc finis K

3 *Livre de divinacions*: see G. W. Coopland, *Nicole Oresme and the Astrologers. A Study of His Livre de divinacions*.

[TABULA HUIUS LIBRI]

f. 31v / Antipodes	.xxx.[1] cap[itu]lo	897 [2]
Arc de la nuit ou nocturne	.xviii.	492
Arc du jour	.xviii.	491
Aux solis	.xxii.	642
Capud Drachonis	.xlvi.	1469
Cauda Drachonis	.xlvi.	1471
Cecle antartique	.xvii.	464
Cecle artique	.xvii.	461
Cenich	.xiii.	368
Centre	.i.	81
Climat	.xxxii.	995
Colure	.xii.	348
Concentrique	.iiii., .xxii.	163 & 637
Conjunction	.xliiii.	1383
Degré	.viii.	270
Dyametre	.i.	83
Eclipse de lune	.xlvi.	1461
Eclipse de soleil	.xlvi., .xlvii.	1457 & 1492
Eclipser	.xlvii., .xlix.	1486 & 1533
Ecliptique	.x.	307, 317
Equidistant	.v.	196
Equinoce	.v., .xviii.	201 & 501

1587 Tabula ... libri] K; *om.* ABCEFGHLMNP 1587-1657 Tabula ... Oresme] *om.* D 1588 capitulo] caplo ECP; *om.* ABFGHKLMN 1589 .xviii.] 28 B 1590 Arc ... 491] *om.* K .xviii.] 28 B 1592 Drachonis] Draconis FHKMN 1593 Drachonis] Draconis FHKMN 1594 .xvii.] .xlvii. GL 1595 .xvii.] .xlvii. GL 1600 .iiii. &] *om.* N & .xxii.] *om.* M 1605 & .xlvii.] *om.* BM; & .xlviii. ACEFGHKLN 1606 & .xlix.] *om.* ABFGHKLMN 1609 .v. &] *om.* KLMN .v. & .xviii.] 28 B; xliii G

1 In mss B, M and P, the chapter numbers are given in Arabic numerals.

2 In the mss these words are linked to the definitions in the text via letters in the margin of the text. Rather than insert many (irrelevant and confusing) letters into the margins of this edition, I have substituted the line numbers to indicate (like Oresme) the exact place of these definitions.

In ms P, only the chapter numbers are given and not the letter-links to specific lines. In ms G, the chapter numbers follow the words but the letter-links precede the words.

Equinocial	.v.	199
Espere	.i.	72
Estade	.xxvi.	760
Estoile fichié	.iii.	136
Excentrique	.xxii.	638
[G]rant cecle	.xvii.	449
Hesmispere	.xiiii.	385
Heure equale	.xxxii.	989
Heure naturel	.xxxii.	990
Instant	.xxi.	588
Jour artificiel	.xviii.	482
Jour naturel	.xviii.	480
Latitude	.xxxiii.	1016
Lever droit	.xix.	569
Lever obliquement	.xix.	564
Longitude	.xxxiii.	1015
f. 32r / Mendre cecle	.xvii.	451
Meridian	.xiii.	366
Milier	.xxxiii.	1019
Minu[te]	.viii.	271
Mouvement journel	.v.	184
[N]adayr	.xlviii.	1508
Opposicion	.xliiii.	1386
Oppositum augis	.xxii.	644
Ozizon	.xiiii.	380
Ozizon droit	.xv.	396
Ozizon oblique	.xv.	396

1615 Grant] rant E .xvii.] 27 B 1617 Heure ... 989] *om.* K 1618 .xxxii.] .xxii. GL 1619 .xxi.] 22 B 1620 .xviii.] *om.* B 1622 Latitude] La latitude G 1624 Lever ... 564] *om.* KP 1629 Minute] Minuit E .viii.] 20 *corr. to* 8 B 1631 Nadayr] adayr E 1632 .xliiii.] .xlviii. KL; 48 B; .xxii. N[3] 1633 .xxii.] .xiiii. N 1634 .xiiii.] .xv. N 1636 .xv.] .xvii. N

3 In ms N, the chapter numbers in lines 1632-1636 are actually the numbers from the following lines. The number in line 2018 has been crossed out, and thereafter the numbers are normal.

[Explicit le traictié que translata en francois tres excellent philozophe Maistre Nicole Oresme.][4]

1637 .xviii.] .xvii. ACEFGHL; ~~.xxviii.~~ N 1648 .i.] 12 B; .xviii. K 1649 .iiii.] 12 *corr. to* 4 B 1652 .xvii.] .xviii. CE; 18 P 1655 .xxvii.] .xxviii. ACEFGHKL; 28 MP; 38 B; *add.* Explicit N 1656-1657 Explicit ... Oresme] HN; *om.* AEGLP; Et finiste la tables des mos F; Explicit C; Explicit le traitier de l'espere K; Explicit tabula vorborum M; [*Illeg.*] libro sit laus et gloria Xristo | Scriptum mense Augustus ano domini 1448 B; *add.* etc. N

4 This explicit, as found in mss H and N – both of which I consider not to be original (see the introduction , p.17) – suggests that the *Traité* is a translation. But it is clear that the Middle French text is not a translation of any earlier text in either Latin or Arabic. This explicit, however, may be the source of later erroneous assertions that it is based on the *De Spera* of Sacrobosco (see the introduction, p. 9).

SOURCE REFERENCES

The following is detailed bibliographic information about the sources cited by Oresme throughout his text.

Cap. I

line 102: Aristotle, *Meteorologica*, Bk. I, Chap. 7: 344a.

Cap. II

line 123: Aristotle, *De caelo*, Bk. I, Chap. 12: 283a25-30.

Cap. IIII

line 172: Aristotle, *De Caelo*, Bk. I, Chap. 9: 279a6-15.

Cap. XXVI

line 756: Johannes de Sacrobosco, *Tractatus de spera*, cap. 1. See Lynn Thorndike, *The Sphere of Sacrobosco and Its Commentators* (Chicago: University of Chicago Press, 1949), pp. 85, 123.

Cap. XXVIII

line 834: Cf. Aristotle, *De caelo*, Bk. II, Chap. 2: 284b6-286a2. (McCarthy erroneously has Bk. II, Chap. 7.)

Cap. XXIX

line 862: Avicenna/Ibn Sīnā, *al-Šifā᾽: al-Ṭabīʻiyyāt, al-Ma'ādin wa-l-āṯār al-ʻulwiyya*, ed. ʻA. al-Ḥalīm Muntaṣir, S. Zāyid, and ʻA.A. Ismāʻīl (Cairo, 1965), p. 26 (ll. 14-27) and pp. 28 (l. 13)-30 (l.15), as cited in Cristina Cerami, "A Map of Averroes' Criticism against Avicenna: *Physics, De caelo, De generatione et corruptione and Meteorology*," in *The Arabic, Hebrew and Latin Reception of Avicenna's Physics and Cosmology*, ed. Dag Nikolaus Hasse and Amos Bertolacci (Berlin, Boston: De Gruyter, 2018), p. 220.

Cap. XXX

line 904: St. Augustine of Hippo, *De Civitate Dei*, Bk. XVI, Cap. 9.

line 923: al-Kindi, *De impressionibus terre et aeris* (also known as *De mutatione temporum*), chap. 1, section 43. See Gerrit Bos and Charles Burnett, *Scientific Weather Forecasting in the Middle Ages: The Writings of Al-Kindi* (London: Kegan Paul, 2000), p. 266.

Cap. *XXXI*

line 951: Aristotle, *Meteorologica*, Bk. II, Cap. 5: 362a33-362b5.

line 966: Pliny the Elder, *Natural History*, Bk. II, Cap. lxviii, 171-175.

Pomponius Mela, *De chorographia*, ed. Piergiorgio Parroni (Rome: Edizioni di Storia e Letteratura, 1984), Bk. I, Cap. 1 (pp. 111-112).

Solinus, *Collectanea rerum memorabilium*, ed. Th. Mommsen (Berlin 1895; rpt. Berlin, 1958), sect. 24, 28 and 40 (pp. 110, 128 and 171).

Priscianus Caesariensis, *Interpretatio ex Dionysio De orbis situ*, in *Prisciani Caesariensis Grammatici Opera*, ed. Augustus Krehl (Leipzig, 1820), 2: 482, ll. 13-15.

Anselm of Laon / Anselmus Laudenensis, "Enarrationes in Evangelium Matthaei," cap. II (*Patrologiae Latinae*, vol. 162, col. 1257B). The ascription of this gloss to Anselm is uncertain, and it is also likely that Ralph of Laon was also (partly) responsible – see *Anselmi Laudunensis Glosae super Iohannem*, ed. Alexander Andrée, Corpus Christianorum Continuatio Mediaevalis 267 (Turnhout: Brepols, 2014), p. xi.

line 967: al-Battānī, *Kitāb al-Zīj*, cap. vi. See Al-Battānī, *Opus Astronomicum*, ed. Carlo Alfonso Nallino (Naples, 1903), 1: 19.

Cap. *XXXII*

line 974: Ptolemy, *Almagest*, Bk. II, Cap. 6.

line 985: Ptolemy, *Tetrabiblos / Quadripartitum*, Bk. II, Cap. 2, sect. 2. (See *Ptolemy's Tetrabiblos in the Translation of William of Moerbeke*, ed. Gudrun Vuillemin-Diem and Carlos Steel [Leuven: Leuven University Press, 2015], p. 199.)

"Haly" is 'Alī ibn Ridwān (388/998-452/1061); his commentary on the *Quadripartitum* has not been critically edited. I have consulted it in conjunction with an early printed version of the *Quadripartitum* – Venice: Bonetus Locatellus, 1493 (USTC no. 991734), fol. 30vb.

line 994: Abū Maᶜšar, *The Great Introduction to Astronomy*, ed. Keiji Yamamoto and Charles Burnett (Leiden: Brill, 2019), Part vi, chap. 2 (1: 603-609); and *The Book of Religions and Dynasties (On the Great Conjunctions),* in Abū Maᶜšar, *On Historical Astrology*, ed. Keiji Yamamoto and Charles Burnett (Leiden: Brill, 2020), 2 vols. The text in *Religion and Dynasties* dealing with the seven climes is found in vol. 1, Appendix 1 ("The Cities and Their Climes, Signs, and Planets"), p. 515; for a discussion of the place the material in this appendix within the works of Abū Maᶜšar, see p. 513.

al-Farghānī, *Jawāmic / Elements*, chap. 8. There are two Latin translations of this text, one by Johannes Hispalensis, which was printed several times under a variety of titles, variations of, for instance, *Astronomorum peritissimi compendium* (Paris: Chrestien Wechel, 1546; USTC 153858) – the material referred to here is on p. 29. The other translation was by Gerard of Cremona and was published in Alfragano, *Il libro dell' Aggregazione delle Stelle*, ed. Romeo Campani (Florence / Città di Castello, 1910), Cap. VIII (p. 90).

line 1006: Abū Maᶜšar, *On Historical Astrology*, vol. 1, Appendix I, p. 515.

Cap. *XXXIII*

line 1021: al-Farghānī, *Jawāmic / Elements*, chap. 8 (= *Astronomorum peritissimi compendium* [Paris, 1546], pp. 27-28; *Il libro dell' Aggregazione*, p. 89).

***Cap.* XXXIIII**

line 1039: al-Farghānī, *Jawāmic / Elements*, chap. 8 (= *Astronomorum peritissimi compendium* [Paris, 1546], pp. 27-28; *Il libro dell' Aggregazione*, p. 89).

line 1049: al-Farghānī, *Jawāmic / Elements*, chap. 8 (= *Astronomorum peritissimi compendium* [Paris, 1546], p. 30; *Il libro dell' Aggregazione*, p. 91).

***Cap.* XXXV**

line 1090: Sacrobosco, *De spera*, cap. 3 (pp. 111-112 and 139-140).

***Cap.* XXXVII**

line 1144: al-Battānī, *Kitāb al-Zīj*, cap. vi. See Al-Battānī, *Opus Astronomicum*, ed. Nallino, 1: 14-20.

***Cap.* XXXVIII**

line 1174: al-Battānī, *Kitāb al-Zīj*, cap. vi. See Al-Battānī, *Opus Astronomicum*, ed. Nallino, 1: 18.

line 1179: Pliny the Elder, *Natural History*, Bk. VI, Cap. xxiv, 89-91. "Canopus" is mentioned in Cap. xxiv, 87 and also in Bk. II, Cap. lxxi, 178,

line 1180a: Solinus, *Collectanea*, sect. 53.1-10 (pp. 195-197).

line 1180b: Abū Maᶜšar, dialogue with his pupil Abū Sādān (9th century) – Abū Saᶜīd Šāḏān ibn Bahr, *Muḏākarāt fī ᶜilm an-nuğūm / Dialogues on the Science of the Stars*, section 10. See Graziella Federico Vescovini, "La versio latina degli *Excerpta de secretis Albumasar di Sadan*," *Archives d'histoire doctrinale et littéraire du moyen âge* 65 (1998), p. 304. Abū Maᶜšar describes Canopus/α Carina as having "moderate light, just as a piece of cloud."

line 1181: The land of Zanğ is also mentioned in Abū Maᶜšar, *The Great Introduction to Astrology*, ed. Yamamoto and Burnett, Part III, cap. 3.7 (1: 251) and cap. 8.13a-8.13b (1: 329-331).

line 1182: Pliny the Elder, *Natural History*, Bk. VI, Cap. xxiv, 81; Solinus, *Collectanea*, sect. 53.2 (p. 196); Pomponius Mela, *De chorographia*, Bk. III, Cap. 7 (sect. 71, p. 167), and comment (p. 421).

lines 1186, 1188: 'Alī ibn Ridwān, commentary on the *Quadripartitum* (Venice: Bonetus Locatellus, 1493), fol. 32rb.

line 1192: Averroës / ibn Rushid, *Commentum magnum super libro De celo et mundo Aristotelis*, ed. Francis James Carmody and Rüdiger Arnzen, Recherches de Théologie et Philosophie médiévales Bibliotheca 4 (Leuven: Peeters, 2003), Bk, II, comm. 111 (vol. 2, pp. 480-481).

line 1200: Aristotle, *De caelo,* Bk. II, Cap. 14: 298a10-15; Averroës, *Commentarium magnum ... De caelo*, Bk. II, comm. 111 (2: 481).

***Cap.* XL**

line 1283: Robert Grosseteste, *Die philosophischen Werke*, ed. Ludwig Baur, Beiträge zur Geschichte der Philosophie des Mittelalters 9 (Munster i. W.: Aschendorff,

1912), sect. X: "De natura locorum," pp. 68-69.

line 1292: Aristotle, *Politics*, Bk. VII, Cap. 13: 1334a32.

Cap. *XLI*

line 1298a: Ptolemy, *Tetrabiblos / Quadripartum*, Bk.II, Cap. 2 (*Translation of William of Moerbeke*, pp. 199-200).

lines 1298b, 1300: 'Alī ibn Ridwān: this could refer to his commentary on the *Quadripartitum* (Venice: Bonetus Locatellus, 1493), fol. 30va-31ra; or to his *Kitab al-bāri' / De iudiciis astororum* – Albohazen Haly filii Abenragel, *Libri de iudiciis astrorum* (Basel: Heinrich Petri, 1551; USTC no. 600067), Bk. VIII, Cap. 36, pp. 405-406: "De potentia planetarum in climatibus et petitione terrae."

line 1305: St. Augustine of Hippo, *De Civitate Dei*, Bk. XVI, Cap. 8.

line 1306: 'Alī ibn Ridwān, commentary on the *Quadripartitum* (Venice: Bonetus Locatellus, 1493), fol. 30va-b.

line 1309: 'Alī ibn Ridwān, commentary on the *Quadripartitum* (Venice: Bonetus Locatellus, 1493), fol. 31ra.

For Hermes, see *Hermès Trismégiste*, ed. Louis Ménard (Paris, 1910), I, ix (p. 50): "Aucune partie du monde n'est vide de démons ...." And Brian P. Copenhaver, *Hermetica*, (Cambridge: Cambridge University Press, 1992), sect. IX (p. 27): "no part of the cosmos is without a demon ...."

Cap. *XLII*

lines 1321, 1325: Aristotle, *Politics*, Bk. VII, Cap. 6: 1327b20-34.

line 1337: Ptolemy, *Tetrabiblos / Quadripartum*, Bk. II, Cap. 2 (*Translation of William of Moerbeke*, pp. 199-200).

line 1340: 'Alī ibn Ridwān, commentary on the *Quadripartitum* (Venice: Bonetus Locatellus, 1493), fol. 31rb.

line 1348: Ptolemy, *Tetrabiblos / Quadripartum*, possibly Bk. II, Cap. 3 (*Translation of William of Moerbeke*, p. 208).

line 1349: 'Alī ibn Ridwān, commentary on the *Quadripartitum* (Venice: Bonetus Locatellus, 1493), fol. 31rb.

line 1358: 'Alī ibn Ridwān, commentary on the *Quadripartitum* (Venice: Bonetus Locatellus, 1493), fol. 31rb.

line 1364: Ptolemy, *Centiloquium*. Oresme is possibly referring to the fourth aphorism: a natural inclination to knowledge is better than having to learn it by study.

Cap. *XLIIII*

line 1374: Averroës, *Commentum magnum ... De celo,* Bk. II, comm. 49 (2: 367-368).

line 1398: Grosseteste, *Die philosophischen Werke*, sect. III: "De sphaera," p. 30.

APPENDIX
MEASUREMENTS

It is always extremely difficult to translate medieval measurements into modern values, although Oresme does at least give a table for his system of measurement in Cap. 26. But to translate this into modern (metric) values is problematic.

In examining both medieval French measures, and the conversions of these into metric values after the French Revolution, it appears that it would be best to begin with the value of the league as given for the region of Beauce, in northern France, between the Seine and the Loire – an area close to Oresme's home and base.

The Beauce league was 3,268 meters.

Oresme's table in Cap. 26 would be as follows:

1 des / finger *or* digit	[2.042 cm]
1 palme [1]/ palm = 4 des	[8.17 cm]
1 pié / foot = 4 palmes	[32.68 cm]
1 pas / stride = 5 piés	[1.634 m]
1 stade / *Lat.* stadium = 125 pas	[204.25 m]
1 millier[2] / mile = 8 stades	[1,634 m][3]
1 leue / league = 2 milliers	[3,268 m]

These metric values, particularly for the smaller units, are fairly close to traditional metric values assigned to them, which makes the choice of the Beauce league for Oresme reasonable.

1 The more traditional "palme" in France was the outstretched "span" from the tip of the thumb to the tip of the fifth or little finger, or about 24.6 cm.

2 Sometimes "mile"/"miles" in Middle French. See line 759.

3 The modern British or American mile is 1609.34 m so Oresme's value is not far off.

APPENDIX
CLIMES

For a detailed and technical survey of the concept of climes (at least up to 1949), see the introduction by Lynn Thorndike to his edition of Sacrobosco's *De spera*, note 88 (pp. 16-18).

1. Ptolemy

Ptolemy does not use the word "clime" in the *Almagest*. Instead, in Book II, cap. 6, he lists 26 parallels which divide the northern hemisphere, each one marking off a latitude where the longest day is one quarter of an hour greater than at the previous line (beginning with the equator where the longest day is 12 hours). See Oresme's Chap. 32, ll. 974-983.

2. Traditional Names

The tradition of naming the seven climes goes back at least as far as Ptolemy where the names are found in Book II, cap. 6 and in the tables at the end of Book II of the *Almagest*, as well as in his *Geography* and *Tetrabiblos*.[1]

The names appear in Sacrobosco near the end of chapter 3[2] and in Oresme's text in Chap. 32, ll. 998-1005.

Clime 1: Diameroës,[3] based on the ancient city of Meroë on the east bank of the Nile, northeast of Khartoum. It is not actually an island (as noted in the text), but the name "Island of Meroë" was used to designate the surrounding area. Latitude: 16°56′ N

Clime 2: Diasyenes, based on the ancient city of Syene (also known as Swenett), now known as Aswan in Upper Egypt. Latitude: 24°05′ N

---

1 Again, see Thorndike's introduction to Sacrobosco's *De spera*, note 88 (pp. 16-18).

2 Sacrobosco, *De spera*, pp. 111-112 and 139-140.

3 The prefix "dia-" is from the Greek preposition διά, meaning "through". There are a number of variations (in Oresme and elsewhere) in the spelling of these names, often beginning with "Dya-" rather than "Dia-".

Clime 3: Dialexandrios, based on the city of Alexandria: Latitude: 31°12′ N

Clime 4: Diarhodos, based on the island (and city) of Rhodes. Latitude: 36°26′ N

Clime 5: Diaromes, based on the city of Rome. Latitude: 41°54′ N

Clime 6: Diaboristenes, based on the ancient Scythian river Borysthenes, now the Dnipro/Dnieper River, emptying into the Black Sea at Kherson. Latitude: ca. 47° N

Clime 7: Diaripheös, based on the Riphean Mountains, a supposed mountain range in north Eurasia, generally in the Volga region. Latitude: ca. 55° N (questionable)

3. Planets and Climes

Planets associated with the climes (Chap. 32, ll. 1006-1012):

CLIME	"PERSIANS"	"GREEKS"
1	Saturn	Saturn
2	Jupiter	Sun
3	Mars	Mercury
4	Sun	Jupiter
5	Venus	Venus
6	Mercury	Moon
7	Moon	Mars

4. Summary of the data given in Chap. 34-36.

Clime		Longest Day	Elevation of the Pole[4]			Width Miles[8] | Leagues	Width Degrees[9]
			Pt[5]	Sa[6]	Or[7]		
	0-1 line	12^h45^m	12°30′	12°45′	12°45′		
1	middle	13^h	16°27′	16°	16°37′30″	440 | 220	7°45′
	1-2 line	13^h15^m	20°14′	20°30′	20°30′		
2	middle	13^h30^m	23°51′	24°15′	24°	400[10] | 200	7°
	2-3 line	13^h45^m	27°12′	27°30′	27°30′		
3	middle	14^h	30°22′	30°45′	30°35′	350 | 175	6°10′
	3-4 line	14^h15^m	33°18′	33°40′	33°40′		
4	middle	14^h30^m	36°	36°24′	36°20′	300 | 150	5°20′
	4-5 line	14^h45^m	38°35′	39°	39°		
5	middle	15^h	40°51′	41°20′	41°15′	255 | 127.5	4°30′
	5-6 line	15^h15^m	43°01′	43°30	43°30′		
6	middle	15^h30^m	45°01′	45°24′	45°22′30″	212 | 105	3°45′
	6-7 line	15^h45^m	46°51′	47°15′	47°15′		
7	middle	16^h	48°32′	48°40′	48°52′30″	185 | 92.5	3°15′
	7 end line	16^h15^m	50°04′	50°30′	50°30′		
Total width of all 7 climes		3^h30^m				2140 | 1070	37°45′

Pt = Ptolemy; Sa = Sacrobosco; Or = Oresme
Longest day at the line between climes from Oresme[11]
Longest day at the "middle" of a clime from Sacrobasco[12]
Note: These figures may contain transmission errors, as copied over time.

4 I.e., the latitude of the various lines defining the climes.
5 Ptolemy, *Almagest*, Book II, cap. 6.
6 Sacrobosco, *De spera*, end of chap. 3 (pp. 111-112 and 139-140).
7 Oresme, chap. 36 (ll. 1115-1128 and ll. 1133-1168).
8 1° = 56⅔ miles. Miles from Sacrobosco, *De spera*, end of chap 3 (pp. 111-112 and 139-140). Leagues (1 league = 2 miles) from Oresme, chap. 34 (ll. 1052-1062).
9 Oresme, chap. 34 (ll. 1054-1062); for the first clime, chap. 36 (ll. 1116-1117).
10 Calculated as 396.66 miles.
11 Oreesme, chap. 35 (ll. 1072-1079)
12 Sacrobosco, *De spere*, end of chap. 3 (pp. 111-112 and 139-140).

5. Oresme's criticism of Sacrobosco (chap. 35)

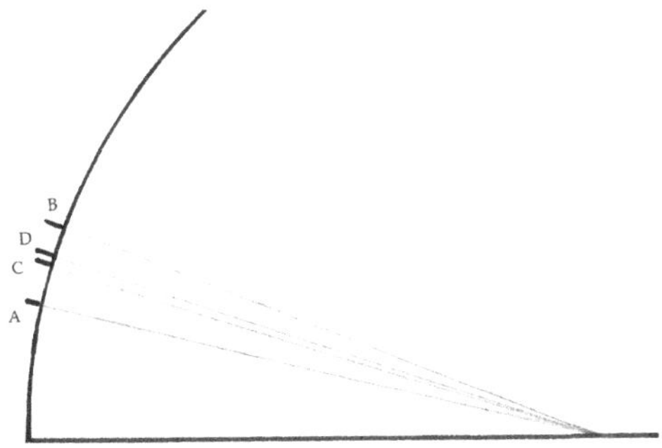

A = beginning of clime 1 (12°45'; longest day = $12^{h}45'$)
B = end of clime 1 (20°30'; longest day = $13^{h}15'$)
C = middle of clime 1 (16°37'30")
D = "middle" longest day of clime 1 (13^{h})

The "middle" longest day (D) is at a latitude higher than the latitude of the middle of the clime (C), since the distance from the beginning of the clime (A) to the "middle" longest day (D) is greater than that of the "middle" longest day (D) to the end of the clime (B), just as the first clime is greater in width than the second clime.

By concentrating only on the "middle" longest day of a clime, Sacrobosco misses this point – that the middle latitude of a clime (C) is always less than the latitude of the "middle" longest day of the clime (D) – and therefore that the length of the longest day at the middle of a clime (in terms of distance) is less than the length of the "middle" longest day (in terms of time).

APPENDIX
PHASES OF THE MOON

The names for the phases of the moon in Diagram 44A do not appear often in medieval texts – probably never in astronomical texts per se, and rarely in cosmological texts. Their appearance here seems to be a one-off presentation by Oresme, without explanation or discussion. The terms do appear in a diagram in an astronomical text by Abbo of Fleury[1] (ca. 945-1004), but his earlier source seems to be Martianus Capella[2] (fl. ca. 410-420). However the Greek terms can be traced back further, at least to Geminos (ca. 1st century BC).[3]

The terms in the diagram and their sources are:

- Early and late crescent moons (often noted in the literature as days 1 and 27 of the lunar month) are given as *monoides* (sometimes *monoydes*), from the Greek, *mēnoeidēs* (μηνοειδής), "crescent shaped".
- The quarter and three-quarter moons (noted as days 8 and 22) are given as *diatomos* (or *dyatomos*), from the Greek, *dichotomos* (διχότομος), "cut in half".
- The moons between the quarter moons and the full moon (noted as days 12 and 18) are given as *amphicirtos* from the Greek, *amphikurtos* (ἀμφίκυρτος), "gibbous".
- The full moon (noted as day 15) is given as *panselenos* from the Greek, *panselēnos* (πανσέληνος), "full moon".

---

1 Ron B. Thomson, "Two Astronomical Tractates of Abbo of Fleury" in J. D. North and J. J. Roche, eds., *The Light of Nature* (Dordrecht: Martinus Nijhoff, 1985), p. 128. Also *Abbonis Floriacensis, Miscellanea de computo, De astronomia et De cosmographia*, ed. Alfred Lohr and Barbara Obrist, Corpus Christianorum Continuatis Mediaevalis 300 (Turnhout: Brepols, 2019), p. 62.

2 Martianus Capella, *On the Marriage of Philology and Mercury*, Book 8 [astronomy], sect. 864.

3 Geminos, *Introduction to the Phenomena*, Book IX, cap. 11-12.

APPENDIX
NEOLOGISMS

Nicole Oresme is well known for having created a number of new French words, both technical terms relevant to the various texts he wrote or translated, and more general terms as well. (Often these are fairly obvious conversions of Latin terms.) Some of these neologisms appear in the *Traitié d'espere*, and these are listed below.

Lillian McCarthy[1] and Menut and Denomy[2] have already developed such lists, but I have re-investigated these words in order to include them here. Part of the problem is that lexicographers have been sparse in their citations, with "14th c., Oresme" being one of their favourites, making it impossible to say that their appearance in the *Traitié* is their first appearance. Even "1361, Oresme" or "1370, Oresme" or "1377, Oresme" does not necessarily mean the *Traitié*, nor does it exclude it as the locus of the first use (sometimes the lexicographers simply use a standard date for any neologism found in any work of Oresme). Some lexicographers have cited Oresme's translation of the *Ethics* of Aristotle, but given the uncertainty of the dating of Oresme's works these words could still have appeared first in our text. In a number of dictionaries no indication is given of the first use of certain words.

I have compared the words in the lists in McCarthy and in Menut and Denomy to entries published in later dictionaries, especially those on line, and I have eliminated some words in their lists because earlier instances (early thirteenth century, late twelfth century) have emerged.

Dictionaries consulted and abbreviations used are:

Bl	Oscar Bloch, *Dictionnaire étymologique de la langue française*, 2 vols. (Paris, 1932)
BW	Oscar Bloch; Walther von Wartburg, *Dictionnaire étymologique de la langue française*, 9th ed. (Paris: Presses Universitaires de France, 1991)
Da_1	Albert Dauzat, et al. *Nouveau dictionnaire étymologique et historique*, 3rd ed. (Paris: Larousse, 1964)
Da_2	Albert Dauzat, *Dictionnaire étymologique de la langue française* (Paris: Larousse, 1938); 10th ed. revised (Paris: Larousse, 1954)

1 McCarthy, pp. 78-90.

2 *Le Livre du ciel et du monde*, pp. 763-773.

DEAF	*Dictionnaire etymologique de l'ancien français* [on-line]
DFSM	*Dictionnaire du français scientifique* [on-line]
DMF	*Dictionnaire de moyen français* [on-line]
Du	Jean Dubois, *Grand dictionnaire étymologique et historique du français* (Paris: Larousse, 2011)
G	Ernst Gamillscheg, *Etymologisches Wörterbuch der französischen Sprache* (Heidelberg, 1928)
GC	Frédéric Godefroy, *Dictionnaire de l'ancienne langue française et de tous ses dialectes du IXe au XVe siècle – Complément*, 3 vols. (Paris, 1893-1902),
HD	Adolphe Hatzfeld, Arsène Darmesteter, and Antoine Thomas, *Dictionnaire générale de la langue française*, 6th ed. 2 vols. (Paris, 1920)
L	É. Littré, *Dictionnaire de la langue française*, 4 vols. (Paris, 1889)

A

altérer, *vb.*

Et si n'y puet estre estoilé ou autre partie adjoustee ou ostee, ne ne puet ... estre corrompu, ne *alteré* en quelconques maniere (lines 115-117)

Most dictionaries cite Oresme (1361, 14th c, or *Ethiques*): B1, BW, Da_1, Da_2, Du, GC, HD, L

DEAF cites Oresme, *Ciel et du monde*, *L'espere*, and *Ethiques*; also de Mondeville (1314)[3] but for medicine only.

DFSM cites de Mondeville (1314), again in terms of medicine only.

antipodes, *subst.*

Et dient que illeuques sont ceulz que l'en appelle "*antipodes*", c'est a dire gens qui ont leurs piés contre les nos, et pour ce que il sont a l'opposite partie de la terre, aussi comme se il fussent sous nous, et nous sous eulx. (lines 897-899)

Most dictionaries cite Corbichon[4] (1372, or 14th c.): BW, Da_1, Da_2, Du, HD

GC and DEAF also cite Oresme, *Ciel et du monde*

Not in DFSM

C

concentrique, *adj.*

Tout cecle qui divise son espere en .ii. moitiés et a son centre ou centre du monde, est dit "*concentrique*". (lines 636-637)

Most dictionaries cite Oresme (1361, or 14th c): B1, BW, Da_1, Da_2, Du, GC, HD, L

---

3 Henri de Mondeville (1260–1320), French surgeon. His *Chirurgie* first appeared in French in 1314.

4 The citation of "1372, Corbichon" is to a text by Jean Corbichon, a French writer and translator, active about the same time as Oresme and also at the court of Charles V. It is possible that the use of a particular neologism by Oresme could precede the use of the word by Corbichon.

DFSM gives Oresme, *L'espere*, as the earliest citation
DMF gives only Oresme,, *Ciel et du monde*

convexe, *adj.*

Et a chascune [espere] une face ou superfice dehors ou dessus qui est appellee "superfice *convexee*". (lines 158-159)

Most dictionaries cite Oresme (1361, 14th c., Oresme, or *Ciel et du monde*): Bl, BW, Da_1, Da_2, Du, G, GC, HD

DFSM gives Oresme, *L'espere*, as the earliest citation

cosmographe, *subst.*

... la porcion dessus dite qui est terre habitable, puet estre devisee en .iii. manieres: en une maniere selonc les hystoriens que l'en appele "*cosmografes*", comme sont Plinius, Pomponius, Solinus, Prescian, Anselme et plusieurs autres ... (lines 964-967)

Most dictionaries cite Oresme (1361, 14c., or Oresme): Da_1, Da_2, DU, GC, HD, L
DEAF gives Oresme, *L'espere*, as the earliest citation
Not in DFSM
DMF gives Oresme, *Politiques*, as the earliest citation

D

démonstration, *subst.*

... duquel je vueil dire en francois generalment et plainement ce qui est couvenable pour savoir a tout homme sans moi profunder es *demonstrations* et es subtiletés qui appartiennent as astrologiens. (lines 12-14)

Most dictionaries cite Oresme (1361, 14th c., or *Ethiques*): B1, BW, Da_1, Da_2, Du, G, GC, HD, L

DFSM gives Nicolas Chuquet,[5] *Géométrie* (1482) as its citation

diffinir, *vb.* (= definir)[6]

Car chascun [mot] est exposé ou *diffini* ou chapistre la ou il est premierement trouvé. (lines 1585-1586)

Most dictionaries which include this meaning give late citations: Da_1, Du: 1425, Chartier;[7] Bl, BW, Da_2, HD, L: 15th c., Villon;[8] G: 15th c.

DEAF is the only dictionary to include "difini" and gives Oresme, *L'espere*, as its citation

E

épaisseur, *subst.*

Et est chascune espere d'une meïsme *espesseur* ou parfondeur en ses parties ... (lines 156-157)

Only one dictionary cites Oresme, 1377: Du; another simply gives 1377: BW; a

---

5 Nicolas Chuquet (ca. 1445/1455–ca. 1488/1500), French mathematician.

6 *Definir* derives from the Latin *definire*, "to limit" or "to end", but it also means "to define". *Diffinir*, which occurs up to the 18th century, derives from the Latin *diffinire* which is a variant of *definire*.

Most dictionaries include only the first meanings ("to end", "to die", etc.) rather than the second ("to define") which is Oresme's meaning here.

7 Alain Chartier (ca. 1385–1430), French poet and political writer.

8 François Villon (ca. 1431–after 1463), French poet.

third gives end of 14th c.: Da
Other dictionaries give examples from the 16th c.: Bl, Da_2, G, HD, L
DEAF does not list this form of the noun under "espoisse".

F

fertile, *adj.*
Item: se les causes especials de bonne habitacion couvenoient et concueroient aveques la general, c'est assavoir terre bien *fertile* et bien assize a souhet ... (lines 1289-1290)
Bl, BW, Da_2 say 14th c.
Other dictionaries (Da_1, Du, GC, HD, L) give later citations
DEAF and DMF give later citations

H

habile, *adj.* (= "propre à")
Et pour ce, dit il que il sont, selonc nature, *habiles* a avoir seignourie sur les autres qui sont plus vers midi ou plus vers septentrion. (lines 1326-1327)
Et par ce appert que ceulx de ces pais de France ou de ces parties ne sont pas communelment *habiles* a savoir astrologie ... (lines 1361-1362)
Most dictionaries cite 14th c., as well as Oresme (*Ethiques*): Bl, BW, Da_1, Da_2, G, GC, HD, L
DEAF cites Oresme; possible 1337 citation as well.
DMF cites Oresme (*Ethiques*, and *Ciel et du monde*)

I

inégal, *adj.*
Item: tous les autres parallelles dessus dis, c'est assavoir tous generalment excepté l'equinocial, l'ozizon oblique les devise en .ii. parties non equales et tant plus est l'ozizon oblique, tant sont ces parties plus inequales. [mss DHKMN = *inegales*] (lines 503-505)
Several dictionaries cite Oresme and his *Éthiques*: Da_1, Du, HD, L; others give citations from later centuries
DEAF cites Oresme, *Éthiques* and *Politiques*
inflexion, *subst.*
Et je cuide que la cause soit pour la lumiere que elle recoit des planetes et des estoiles, et de la reflexion [ms F = *inflexion*] de l'air, comme dit est ... (lines 1542-1543)
Several dictionaries cite 14th c. and 1390, E. du Conty:[9] Bl, BW, Da_1, GC, HD
Not in DEAF; DMF refers to GC
instant, *subst.*
... il avient chascun jour naturel une fois par le mouvement journel du zodiaque et de ses poles que le pole du zodiaque est aveques leur cenith ... par .i. seul *instant*. Et est *instant* .i. moment de temps indivisible qui est tantost passé et

9 Évard de Conty (late 14th c.–1405), French court doctor and writer. Assumed to have been known to Oresme.

soudainement. (lines 585-589)

Several dictionaries cite 14th c. and 1377 Oresme: BW, Du: Da_1, Du, G; others cite 16th c.: Bl, Da_2, GC, HD, L

DEAF cites Oresme, *L'espere* and *Ciel et du monde*

intersequer, *vb.*

Et cest cecle ici traverse et *interseque* l'equinocial en .ii. poins contraires ... (line 247)

Ceuls ont droit ozizon qui habitent dessous l'equinocial, qui ont leur cenith en l'equinocial, car leur ozizon *interseque* et divise l'equinocial en croiz et a droit angles. (lines 396-398)

DEAF cites Oresme, *L'espere* and *Ciel et du monde et*

J

judicative, *adj.*

Et pour ce ... il appert que ceulz de ce pais ne pueent pas communelment profiter en astrologie *judicative* ... (lines 1365-1366)

Godefroy[10] cites Oresme (*Éthiques*); L: 14th c., Oresme

DEAF cites Oresme, *L'espere*

L

latitude, *subst.* (geographical, astronomical meaning)

De la *latitude* du zodiaque. (line 301)

Chascun climat ha longitude et *latitude*, c'est a dire longueur de lay. (line 1014)

De la *latitude* des climats. (line 1036)

Many dictionaries give earlier citations for "latitude" meaning "width", as opposed to "latitude" as related to "longitude"

Various dictionaries cite Oresme: B1, BW, Da_1, Da_2; others cite later instances.

DEAF cites Oresme, *Politiques*

DMF cites Oresme, *Ciel et du monde*

longitude, *subst.* (geographical, astronomical meaning)

Chascun climat ha *longitude* et latitude, c'est a dire longueur de lay. (line 1014)

Earlier uses of "longitude" refer to "length".

Many dictionaries cite Oresme (1361. 1377, 14th c.): B1, BW, Da_1, Du, GC, HD, L

DEAF cites Oresme, *L'espere* and *Ciel et du monde*

DMF cites Oresme, *Ciel et du monde*

M

meridian, *subst.* (great circle through the poles of the earth)

Un autre cecle est en l'espere materiel et ymaginé ou ciel qui est appellé "*meridian*", et passe par les poles du monde et par le point du ciel qui est tout droit sus nostre teste (line 366-368)

BW, Da_1 cite "1377, Oresme"

DEAF and DMF both cite Oresme, *Ciel et du monde*. Other (earlier) citations possibly have other meanings (e.g., "southern").

---

10 Frédéric Godefroy, *Dictionnaire de l'ancienne langue française* ,,, (Paris, 1885).

modéré, *adj.*

... quant a l'abitacion de la terre, une cause general est au regart du ciel, c'est asavoir *moderee* ou attrempee distance de la voie du soleil ... (lines 1265-1266)

Some dictionaries cite Oresme (1361, 1370, 14th c., *Éthiques*): Bl, BW, Da_1, Da2, Du, L; or just 14th c: G, HD

DEAF cites Oresme, *L'espere*, *Ethiques*, and *Politiques*

DMF cites Oresme, *Ethiques*, and also Jean Daudin[11] (ca. 1370) and Jacques de Hemricourt[12] (1360-1399)

N

nadair, nadir, *subst.*[13]

Pour ce que la terre est moult plus petite que le soleil, son umbre procede derriere soy.... Et est tousjours la pointe ... de celle umbre a l'opposite du soleil est appellé *"nadair* solis". (lines 1505-1508)

Some dictionaries cite Oresme (1361, 1370, 14th c.): Bl, BW, Da_1, Da_2, Du, GC, HD, L

DEAF only links to other dictionaries.

DMF cites Oresme, *L'espere* and *Ciel et du monde.*

O

opaque, *adj.*

Et [les planettes] ne sont pas corps *opacques* ou obscurs tant comme est la lune ... (lines 1445-1446)

All dictionaries, including DEAF and DMF, cite first entries as 15th or 16th century[14]

DEAF cites Meunier, *Essai*, p. 65, but there is nothing on that page, which actually deals with his treatise on money.

P

palme, *subst.*

... et .i. pié vault .iiii. *palmes*, et une *palme* .iiii. des. (lines 760-761)

A few dictionaries (Bl, BW, Da2) cite "14th c. Oresme"; one (Da_1) cites " mid- 12th c."

---

11 Vincent of Beauvais, *De eruditione filiorum nobilium*, trans. Jean Daudin, ca. 1370.

12 Jacques de Hemricourt (fl. 1360-1399), *Le patron de la temporalité*.

13 Most dictionaries note that Oresme uses the form "nador" although I have not found it so written in this text. DMF cites "nadair"; Meunier (*Essai*, p. 189) cites "nadoir". Oresme's phrase is something of an odd construction, since "nadir" comes from the Arabic, and "solis" is the Latin genitive.

14 Several dictionaries do refer to the 14th century, but they cite the *Complainte de Nature à l'Alchimiste errant* which was long attributed to Jean de Meun/Meung (c. 1240–c. 1305). However this text is actually from the 16th century and is by Jean Perréal of Paris (ca. 1450/1460–1530). See André Vernet, "Jean Perréal et la *Complainte de Nature* attribuée à Jean de Meun," *Comptes rendus des séances de l'Académie des Inscriptions et Belles-Lettres*, année 87 (1943), pp. 93-100.

DEAF gives many citations, but the dating is not obvious.

DMF cites Jean Froissart[15] (1373), Philippe de Mezières[16] (ca. 1386-1389), and Jean d'Arras[17] (c. 1392-1393)

plage, *subst.*

Et est assavoir que selonc ces .v. zones sont en la terre .v. parties ou *plagez* ou regions desquelles chascune est sous aucunes de ces .v. zones du ciel dessus dites. (lines 822-824)

Most dictionaries cite examples from the 15th and 16th centuries.

DEAF does not have this meaning.

DMF cites Oresme, *Ciel et du monde*

policie, *subst.*

Et cuide que aveques la cause dessus dite de maintenir roial seignourie et bonne *policie*, sont autrez causes ... (lines 1329-1330)

Some dictionaries cite Oresme (1361, 14th c. *Ethiques*): Da_1: L; others give simply 1361, 1370 or 14th c.: Bl, BW, Da_2, Du

DEAF cites Oresme, *Traité des monnaies*

DMF cites Oresme, *Livre de divinaction, Livre de Ethique, Traité des monnaies*, and *L'espere*

R

récapituler, *vb.*

Je di donques, en *recapitulant* par ordre retrograde ... (line 626)

Most dictionaries cite Oresme (1361, 1370, 14th c., or *Ethiques*) Bl, BW, Da_1, Da_2, Du, HD, L

DEAF cites Oresme, *L'espere, Politiques,* and *Ethiques*

DMF cites *Politiques, Ethiques,* and *Ciel et du monde.*

refléter, *vb.*

... pour laquelle chose l'espere du feu et l'espere de l'air recoivent lumiere du soleil et la retournent ou *reflectent* ... (lines 1535-1536)

Some dictionaries cite Oresme (1361, 14th c.): Da_1: HD, L:

DEAF cites – for *reflechir* – Jean de Meun's 1284 translation of Flavius Vegetius Renatus, *De re militari.*

DMF cites – for *reflectir* – Évard de Conty

réflexion, *subst.*

Car lors [le soleil] giete ses rays plus droit sur la terre, et fait plus droite *reflexion* de lumiere ... (lines 918-919)

Several dictionaries cite Oresme, 14th c.: Bl, BW, Da_2, HD, L

DEAF cites Oresme, *L'espere* and *Ciel et du monde*

DMF cites Corbechon and Oresme, *Ciel et du monde*

---

15 Jean Froissart (c. 1327– c. 1405), French writer. The text cited is *Le Joli buisson de jonece.*

16 Philippe de Mezières (1327-1405), French author. The text cited is *Le Songe du vieil pelerin.*

17 Jean d'Arras (fl. 1392-1394), French writer. The text cited is *Mélusine.*

répondre, *vb.* (= "correspond")

... auquel ozizon il *respont* .ii. cecles en l'espere materiel. (lines 448-449)

Several dictionaries cite 14th c. sources (but with this meaning?)

Neither DEAF nor DMF give examples with this meaning.

S

semidiametre, *subst.*

... et chascune droite ligne qui passe au travers de la terre, ou de quelconqucs espere ou cercle, parmi le centre de cele espere ou de ce cercle est appelée "dyametre" de celle chose; et la moitié du dyametre qui part du centre est appellé *semidyametre*. (lines 81-84)

Not listed in most dictionaries – most simply indicate that words with the prefix "semi-" are equivalent to words with the prefix "demi-".

DEAF cites Oresme, *L'espere* and *Ciel et du monde*

DMF cites Oresme, *Ciel et du monde*

stérile, *adj.*

... l'autre est quant elle est *sterile* et ne porte bon fruit ... (lines 1271-1272)

Several dictionaries cite Oresme, especially the *Ethiques*: BW, Da_1, Du, HD,

DEAF cites Oresme.

DMF cietes Oresme, *Ethiques*

T

traitié / traité, *subst.*

Et se ai je declairié plus plainement en .i. *traitié* que je ai fait de ceste matiere. (lines 1367-1368)

Some dictionaries cite Oresme (1361, 1370, 14th c): Bl, BW, Da_1, Da

G mentions a 13th c. text but gives no details.

DEAF includes Oresme among its earliest citations – *L'espere* and *Livre de divinacions*

DMF also includes Oresme among its earliest citations – *Ethiques* and *Ciel et du monde*

tropique, *subst.*

Les .ii. poins de ces .ii. solstices par leur mouvement journel descripsent .ii. des cecles dessus dis qui sont appellés "*tropiques*"... (lines 455-456)

A few dictionaries mention Oresme (1377) – BW, Da_1 – but many give citations from the 16th c.

DEAF cites Oresme (*L'espere, Politiques*, and *Ciel et du monde*)

DMF cites Oresme (*Politiques,* and *Livres du ciel*)

BIBLIOGRAPHY

Primary sources

Primary sources which appear in multiple editions and multiple languages (e.g., the works of Aristotle or Augustine of Hippo) are listed by author and title only.

Abbo of Fleury. *Abbonis Floriacensis, Miscellanea de computo, De astronomia et De cosmographia*. ed. Alfred Lohr and Barbara Obrist. Corpus Christianorum Continuatis Mediaevalis 300. Turnhout: Brepols, 2019).

Abū Maᶜšar. *The Great Introduction to Astronomy*. Ed. Keiji Yamamoto and Charles Burnett. Leiden: Brill, 2019.

—. *The Book of Religions and Dynasties (On the Great Conjunctions)*. In Abū Maᶜšar, *On Historical Astrology*, ed. Keiji Yamamoto and Charles Burnett. 2 vols. (Leiden: Brill, 2020.

Abū Saᶜīd Šāḏān ibn Bahr. *Muḏākarāt fī ᶜilm an-nuǧūm / Dialogues on the Science of the Stars*. See Graziella Federico Vescovini, ed. "La versio latina degli *Excerpta de secretis Albumasar di Sadan." Archives d'histoire doctrinale et littéraire du moyen âge*, 65 (1998), 271-330.

'Alī ibn Ridwān. Commentary on the *Quadripartitum*. Published in conjunction with Ptolemy, *Quadripartitum*. Venice: Bonetus Locatellus, 1493 (USTC no. 991734)

—. *Kitab al-bārᶜ / De iudiciis astororum* – Albohazen Haly filii Abenragel, *Libri de iudiciis astrorum*. Basel: Heinrich Petri, 1551 (USTC no. 600067)

Anselm of Laon / Anselmus Laudenensis, "Enarrationes in Evangelium Matthaei," In *Patrologiae Latinae*, vol. 162, cols. 1227-1500.

—. *Anselmi Laudunensis Glosae super Iohannem*. Ed. Alexander Andrée. Corpus Christianorum Continuatio Mediaevalis 267. Turnhout: Brepols, 2014.

Aristotle. *De Caelo*.

—. *Metaphysica*.

—. *Meteorologica*.

—. *Politics*.

Augustine of Hippo. *De Civitate Dei / City of God*.

Averroës / ibn Rushid. *Commentum magnum super libro De celo et mundo Aristotelis*. Ed. Francis James Carmody and Rüdiger Arnzen. Recherches de Théologie et Philosophie médiévales Bibliotheca 4. Leuven: Peeters, 2003.

Avicenna / ibn Sīnā. *Al-Šifāʾ: al-Ṭabīʿiyyāt, al-Maʿādin wa-l-āṯār al-ʿulwiyya*. Ed. ʿA. al-Ḥalīm Muntaṣir, S. Zāyid and ʿA.A. Ismāʿīl. Cairo, 1965.

al-Battānī. *Kitāb al-Zīj* In Al-Battānī, *Opus Astronomicum*, ed. Carlo Alfonso Nallino. 3 vols. Naples, 1899-1907.

al-Fraghānī, *Jawāmic / Elements*: Trans. Johannes Hispalensis (printed several times under a variety of titles, for instance, *Astronomorum peritissimi compendium*. [Paris: Chrestien Wechel, 1546; USTC 153858]).

—. —: Trans. by Gerard of Cremona. In Alfragano, *Il libro dell' Aggregazione delle Stelle*, ed. Romeo Campani. Florence/Città di Castello, 1910.

Geminos. *Introduction to the Phenomena*.

Grosseteste, Robert. *Die philosophischen Werke*. Ed. Ludwig Baur. Beiträge zur Geschichte der Philosophie des Mittelalters 9. Munster i. W.: Aschendorff, 1912.

Hermes. *Hermès Trismégiste*. Ed. Louis Ménard. Paris, 1910.

—. *Hermetica*. Ed. Brian P. Copenhaver. Cambridge: Cambridge University Press, 1992).

Herodotus. *History*.

Hesiod. *Works and Days*.

ibn Rushid. *See* Averroës.

ibn Sīnā. *See* Avicenna.

al-Kindi. *De impressionibus terre et aeris* (also known as *De mutatione temporum*). In Gerrit Bos and Charles Burnett, *Scientific Weather Forecasting in the Middle Ages: The Writings of Al-Kindi*. Pp. 263-310. London: Kegan Paul, 2000.

Martianus Capella. *The Marriage of Philology and Mercury*.

Oresme, Nicole. *Nicole Oresme, De proportionibus proportionum and Ad pauca respicientes*. Ed. Edward Grant. Madison: University of Wisconsin Press, 1966.

—. *Nicolai Oresme. Expositio et Quaestiones in Aristotelis De anima*. Ed. Benoît Patar. Louvain-la-Neuve: Éditions de l'Institut Supérieur de Philosophie, Louvain-Paris: Éditions Peeters, 1995.

—. *Livre de divinacions*. Ed. G. W. Coopland, *Nicole Oresme and the Astrologers. A Study of His Livre de divinacions*. Cambridge MA: Harvard University Press, 1952.

—. *Le Livre du ciel et du monde*. Ed. Albert D. Menut and Alexander J. Denomy. Madison: University of Wisconsin Press, 1968.

—. *Questiones super De celo*: Claudia Kren, ed. "The Questiones super *De celo* of Nicole Oresme." Doctor of Philosophy thesis, University of Wiesconsin, 1965.

—. *Questiones super De spera*: Garrett Droppers, ed. "The Questiones De sphera of Nicole Qresme." Doctor of Philosophy thesis, University of Wisconsin, 1966.

—. —: Alain Boureau, ed. *Nicole Oresme, Deux suites de questions sur la sphère ....* Astronomie, dioptrique et cinétique tome x. Paris: Les Belles Lettres, 2024.

—. *Questiones super De spera* (Erfurt ms): Alain Boureau, ed. *Nicole Oresme, Deux suites de questions sur la sphère ....* Astronomie, dioptrique et cinétique tome x. Paris: Les Belles Lettres, 2024.

—. *Traitié de l'espere*: Lillian McCarthy, ed. "Maistre Nicole Oresme, Traitié de l'Espere." Doctor of Philosophy thesis, University of Toronto, 1943.

—. —: Terrence J. McGovern, ed. "Traité de l'Espere and Minor Astrological Works: A Critical Edition." Doctor of Philosophy thesis, The Catholic University of America, 1974.

—.—: John Vernon Myers, ed. "Maistre Nicole Oresme, Le Traité de la Sphère." Master of Arts thesis, Department of Romance Languages, Syracuse University, 1940.

—. —: Johannes de Sacro Bosco. *Le traicté de la sphere*, translaté de latin en françois par maistre Nicole Oresme. Paris: Chez Simon du Bois pour Regnault Chaudière, [1528]. (USTC 8403–this is actually Oresme's text and not Sacrobosco's)
—. [For a list of other works by Oresme, see Chapter 1, p. 2 note]

Phares, Simon de. *Le* Recueil des plus celebres astrologues *de Simon de Phares*. Ed. Jean-Patrice Boudet. 2 vols. Paris: Champion, 1997-1999.
Pliny the Elder. *Historia Naturalis / Natural History*
Pomponius Mela. *De chorographia*. Ed. Piergiorgio Parroni. Rome: Edizioni di Storia e Letteratura, 1984.
Priscianus Caesariensis. *Interpretatio ex Dionysio De orbis situ*. In *Prisciani Caesariensis Grammatici Opera*, ed. Augustus Krehl. 2: 478-508. Leipzig, 1820.
Ptolemy. *Almagest*.
—. *Tetrabiblos / Quadripartitum*. In *Ptolemy's Tetrabiblos in the Translation of William of Moerbeke*, ed. Gudrun Vuillemin-Diem and Carlos Steel. [Leuven: Leuven University Press, 2015.

Sacrobosco, Johannes de. *Tractatus de spera*. Ed. Lynn Thorndike, *The Sphere of Sacrobosco and Its Commentators*. Chicago: University of Chicago Press, 1949).
—. See also Oresme, Nicole. *Traitie de l'espere* ....
Solinus, Julius. *Collectanea rerum memorabilium*. Ed. Th. Mommsen. Berlin 1895; rpt. Berlin: Wiedmann, 1958.

Secondary Sources

Bloch, Oscar. *Dictionnaire étymologique de la langue française*, 2 vols. Paris, 1932.
—, and Walther von Wartburg. *Dictionnaire étymologique de la langue française*. 9th ed. Paris: Presses Universitaires de France, 1991.

Cerami, Cristina. "A Map of Averroes' Criticism against Avicenna: *Physics, De caelo, De generatione et corruptione and Meteorology*." In *The Arabic, Hebrew and Latin Reception of Avicenna's Physics and Cosmology*, ed. Dag Nikolaus Hasse and Amos Bertolacci. Pp. 163-240. Berlin, Boston: De Gruyter, 2018,.
Clagett, Marshall. "Oresme, Nicole." *Dictionary of Scientific Biography*, ed. Charles C. Gillispie. 10: 223-230. New York: Charles Scribner's Sons, 1974.
"Compte Rendu de la Séance tenue a Paris les 18 et 19 octobre 1925": "Establissement de règles pratiques pour l'édition des anciens textes français et provençaux", *Romania*, 52 (1926), 243-249.
Courtenay, William J. "The Early Career of Nicole Oresme." *Isis* 91 (2000), 542-548.

Dauzat, Albert. *Dictionnaire étymologique de la langue française*. Paris: Larousse, 1938. 10th ed. revised, Paris: Larousse, 1954.
—, et al. *Nouveau dictionnaire étymologique et historique*, 3rd ed. Paris: Larousse, 1964.
Dictionnaire de moyen français [on-line]
Dictionnaire du français scientifique [on-line]

Dictionnaire etymologique de l'ancien français [on-line]

Di Liscia, Daniel A. and Aurora Panzica. "The Writings of Nicole Oresme: A Systematic Inventory." *Traditio* 77 (2022), 235-375

Dubois, Jean. *Grand dictionnaire étymologique et historique du français*. Paris: Larousse, 2011.

Duhem, Pierre. *Le système du monde*. 10 vols. Paris, 1913-1959.

Fery-Hue, Françoise. "Le mathématicien Nicolas Chuquet et la planche à régler." *Pecia* 13 (2010) [= *Du scriptorium à l'atelier*], pp. 329-344.

Gamillscheg, Ernst. *Etymologisches Wörterbuch der französischen Sprache*. Heidelberg, 1928.

Godefroy, Fréderic. *Dictionnaire de l'ancienne langue française et de tous ses dialectes du IXe au XVe siècle*. 8 vols. Paris, 1881-1893.

—. —: – *Complément*, 3 vols. Paris, 1893-1902.

Gumbert, J.P. "Medieval Manuscripts in French in the Leiden University Library: A Handlist." In Peter Rolfe Monks and D.D.R. Owen, eds., *Medieval Codicology, Iconography, and Translation*. Studies for Keith Val Sinclair, pp. 28-47. Leiden: Brill, 1994.

Hagen, Hermann. *Catalogus codicum Bernensium*. Bern, 1874.

Hanna, Ralph, and Jeremy Griffiths. *A Descriptive Catalogue of the Western Medieval Manuscripts of St John's College Oxford*. Oxford: University Press, 2002).

Hatzfeld, Adolphe, Arsène Darmesteter, and Antoine Thomas. *Dictionnaire générale de la langue française*. 6th ed. 2 vols. Paris, 1920).

Jourdain, Charles. "Nicole Oresme et les astologues à la cour de Charles V." *Revue des Questions Historiques*, 18 (1875), 136-159. {Pamphlet 10223}

Laird, Edgar. "Astrology in the Court of Charles V of France, as Reflected in Oxford, St John's College, ms 164." *Manuscripta* 34 (1990), 167-176.

Langlois, Ernest. "Notices des manuscrits français et provençaux de Rome antérieurs au xvie siècle." *Notices et Extraits des Manuscrits de la Bibliothèqe Nationale et Autres Bibliothèques*, 33 part 2 (1889), 1-347

Littré, É. *Dictionnaire de la langue française*. 4 vols. Paris, 1889.

Menut, A. D., and A. J. Denomy, "Maistre Nicole Oresme, Le Livre du ciel et du monde," *Mediaeval Studies* 5 (1943), 167-333.

Meunier, Francis. *Essai sur la vie et les ouvrages de Nicole Oresme*. Paris, 1857.

Moreau, Brigitte. *Inventaire chronologique des éditions parisiennes du xvie siècle*. Vol. 3. Abbeville: F. Paillart,1985.

Omont, Henri. *Bibliothèque nationale. Catalogue général des manuscrits français. Nouvelles acquisitions françaises, 4: nos 10001-11353 et 20001-22811*. Paris, 1918.

Ouy, Gilbert. "Jean Lebégue (1368-1457), auteur, copiste et bibliophile." In Godfried Croenen and Peter Ainsworth, eds., *Patrons, Authors and Workshops*, pp. 143-172. Louvain: Peters, 2006.

Shaw, D.J. "New Dates in the Career of Simon Du Bois, Reformation Printer, Paris 1523-1529." *Yale University Library Gazette*, 67 (1992-1993), 32-36.

Sinner, J. R. *Catalogus codicum mss. bibliothecae Bernensis*. Bern, 1760.

Thomson, Ron B. "Two Astronomical Tractates of Abbo of Fleury." In J. D. North and J. J. Roche, eds., *The Light of Nature*, pp. 113-133. Dordrecht: Martinus Nijhoff, 1985.

Verne, Jules. *Le Tour du monde en quatre-vingts jours / Around the World in Eighty Days*. 1872.

Vernet, André. "Jean Perréal et la *Complainte de Nature* attribuée à Jean de Meun." *Comptes rendus des séances de l'Académie des Inscriptions et Belles-Lettres*, année 87 (1943), 93-100.

Wright, John Kirland. *The Geographical Lore of the Time of the Crusades*. New York: American Geographical Society, 1925 / Dover Reprint, 1995.

INDEXES

To access all references, the reader should consult the topic in both indexes – General and Text.

General Index

The General Index is to page numbers. The text itself has a separate index tied to line numbers; however, entries from the footnotes and comments to the text are listed here, by page number.

Text Index

The index to the text is by line number. The source references are listed in the General Index. Where there are references to diagrams, and there are two versions for the same chapter, they are noted as, e.g., "Diag. 8a", or "Diag. 8b", or as "Diag. 8a-b" (where something appears in both).